Starch derivatization

Starch derivatization

Fascinating and unique industrial opportunities

K.F. Gotlieb
A. Capelle

Wageningen Academic
P u b l i s h e r s

Subject headings:
Granular structure
Aqueous alcohol
Spacers
New starch esters

ISBN 9076998604

First published, 2005

Wageningen Academic Publishers
The Netherlands, 2005

Preface

As Whistler mentions in his great book on starch technology (1) the use of starch both for food and industrial purposes is millennia old. Modern starch chemistry dates from the late 1950's and is characterized by the conversion of native starch into derivatives. Early it was discovered that for a successful application starch had to be modified. Besides the classical applications of starch in food products, the paper- and textile industry and the use as adhesives, new applications as super absorbents and bioplastics have been developed.

Starch is produced industrially on a large scale from various plants such as corn, tapioca and potato. It is renewable, "green" and cheap. Although the starches from these plants differ, the primary applications areas are the same. As a result of the technological advances in the genetic modification of plants one can expect that in the coming years the differences in starches from different plants will disappear and that complete new types of starches will be developed.

Currently, a relatively small number of chemicals are used for the modification of starch. By varying the degree of modification and the combination of different modifications a large number of products can be prepared. New possibilities are created by the use of new chemicals and the preparation of highly substituted products in non conventional solvents such as aqueous alcohol. Using newly developed enzymes also increases the scope of possibilities. The increase in knowledge about the structure of the starch granule presents possibilities for specific modifications. Today an important goal of the starch industry is to create products that are sustainable thereby reducing reliance on non renewable resources.

This book describes a number of new developments and addresses topics such as:
- relation between granular structure and derivatization;
- derivatization in aqueous alcohol;
- new starch esters;
- use of spacers in derivatization;
- enzymatic derivatization.

This book is intended for the industrial starch chemist as well as anyone else interested in the derivatization and application of polysaccharides and natural polymers.

Without the help of Jos Meiberg and Sergei Pogrebnoi the realization of this book would have been very difficult. Their assistance is much appreciated.

It presents ideas and incentives for the development and commercialization of non traditional new carbohydrate derivatives.

This book is dedicated to W.C.Bus, our mentor and friend.

References

1. R.L. Whistler, J.N. BeMiller & E.F. Paschali, Starch, Chemistry and Technology. 1984; Academic Press, London.

Contents

The starch granule, effects of crystallinity and surface on derivatization

Characteristic for a specific starch are the shape, size and size distribution of the starch granules and the detailed structure of the granule such as the structure of amylose and amylopectin. The crystalline structure of the starch granule and hence the X-ray spectrum is determined by the chain length of the amylopectin molecules which are part of the so called cluster-structure. A regular ordering of double helixes results in a structure as presented in Figure 1.

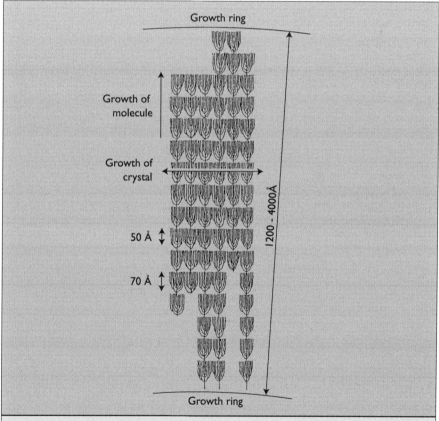

Figure 1. Schematic representation of the arrangement of amylopectin molecules within a growth ring (from French, 1982).

Most root- and tuber derived starches like potato starch are characterized by a B-type X-ray spectrum and grain derived starches like corn by an A-type spectrum. Native starch granules are semi-crystalline, potato starch granules are 28% crystalline and corn starch granules 40%. A relationship exists between the structure of the starch granule and the end result of the derivatization of granular starch. Even during the derivatization of partly gelatinized starch the original granular structure is important.

Structural elements, surface layers

Oostergetel and van Bruggen (1) have shown that the amylopectin molecules are part of a super-helix (Figure 2). This "Oostergetel and van Bruggen" model was confirmed by small angle X-ray scattering of the starch granule (2). A recent review of Gallant et al. (3) introduces (again) "blocklets" as structural element

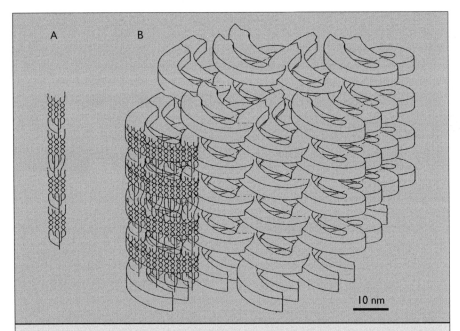

Figure 2. A three-dimensional model of the structure of starch (1). (A) The lineair chains in a cluster form double helices. (B) Double helices of adjacent amylopectin molecules form a more or less continuous left-handed helical crystal. The crystalline superhelices are arranged in a tetragonal array.

of the granule. These are small globular particles with, depending on the biological source, a diameter of 20-500 nm (potato starch 200-500 nm).

Use of NMR techniques has increased our knowledge about the distribution of water in the starch granule (4). Jay-lin Jane et al. (5, 6) peeled starch granules with aqueous solutions of 4M $CaCl_2$ or 13 M LiCl. As gelatinization in these solutions occurs from the outside to the inside it is possible after removal of the gelatinized material to determine the composition of the surface layers. Jay-lin Jane concluded:
- Amylose is predominant on the outside of the granule.
- Amylose in the center of the granule has a higher M_w than amylose on the outside.
- Amylopectin in the center has longer B-chains.
- Phosphate groups in potato starch are concentrated in the center of the granule.

In an other study about the morphological changes of different starches by surface gelatinization with aqueous $CaCl_2$ this researcher concluded (7) that the various starches can be subdivided into three groups:
- starches with regular gelatinized surfaces (potato, waxy potato, sweet potato, corn and high-amylose corn);
- starches with gelatinization at specific spots (equatorial groove) (wheat, barley and high amylose barley;
- starches who after gelatinization cannot be separated into individual granules (waxy barley, waxy corn and rice).

The same researcher investigated $POCl_3$ cross-linked starches with GPC and ^{31}P NMR and concluded that the amylose molecules are randomly dispersed in the starch granule instead of in bundles (8).
Using enzyme-gold labeling indicated that there exists alternating layers of dense packed amylopectin and amylose molecules. The outer layer of the granule consists of amylopectin, in the center there is also amylopectin (9).

The previous paragraph suggests that a starch granule is certainly not homogeneous in terms of composition and structure. A certain layering is also shown by the existence of growth rings, visible microscopically, especially in large (potato) starch granules. These growth rings are probably the result of rhythmic physiological fluctuations in the potato starch plant causing variations in the carbohydrate supply for starch deposition. The swelling of starch granules sets the afore mentioned spherical particles ("blocklets") free from the growth ring structure (10, 11). It might be possible that the "starch molecules" end at the border of the ring and that the size of the molecules is more or less equal to the size of the ring.

The layered structure indicates the probability of differences in properties between the outer layer(s) and the other parts of the starch granule.

Galliard (12) mentions:
"Despite the fact that the outer surface of granules must play an important part in many applications of starch there is surprising lack of definite information on the nature of the surface. Do the polysaccharide chains extend to form the limiting boundary of the granule or are there non-starchy material, wholly or partly covering the surface? If so how does this influence properties such as surface charge, penetration of water, ions and small or large molecules? α-amylases appear to attach to different types of granules in different ways: what factors control the binding of enzymes to starch granules?" and he continues *"so what does small molecule "see "as it approaches an isolated starch granule? Does it encounter a solid boundary or a network of pores, which pass through a whole structure of the granule? If this is the latter, does this mean that there can be surfaces within the granule"*

This statement is still surprisingly actual today, as shown by the publication of J.N. BeMiller (13) which stresses the importance of knowing the structure of granules in view of new ways and possibilities for starch derivatization. The same author mentions in other publications (14, 15, 16) the possible influence of pores, channels and cavities on the penetration of chemicals in the starch granule. It can be generally concluded that the reactivity of the starch granule not only is determined by the outer granule surface but also by the inner surfaces.

Channels, pores and voids

As probably channels and pores do not occur in potato starch granules, reactants diffuse very slowly through the outer surface. The penetration of fluorescent fatty-amides as model substance for reagents into waxy corn, corn and potato starch under different circumstances was microscopically followed (17). The fatty amides were prepared from fatty acid chlorides (C_6 - C_{16}) and aminopyridine.

Some conclusions:

- The researched starches differ in the penetration of the compounds into the granule at room temperature. Corn starch granules cannot be penetrated by molecules with a carbon chain length longer than C_{14}, potato starch granules not by molecules with a carbon chain length longer than C_{12}.
- swelling inhibitors change the penetration of the studied compounds under etherifying conditions. Sodium citrate has more effect than sodium sulfate.
- The area around the hilum is better penetrated than the rest of the granule.
- The outer layer of the starch granule is less penetrated than the surface inside channels and cavities and the starch matrix.
- In case of damaged starch granules, an accumulation of the fluorescent occurs in the center of the granule besides a thick non colored layer at the outer surface.

Currently the general assumption is that the outer layer of a starch granule has a specific structure and composition. This assumption is used to explain various phenomena. Investigation into the partition of nitrogen in granular cationic starch with ESCA showed that the cationization occurred faster into the inner direction than on the outer shell. This is explained on the basis of the structure of the granule (18). The starch granule possesses a dense outer layer consisting mainly of crystallized amylopectin molecules. This layer hinders the penetration of the reagent, however 'microchannels" exist which promote the penetration.

The diffusion velocity could depend on the concentration gradient of the reagent in the vicinity of the granule. The inside of the granule could contain a larger proportion of amorphous material and hence give a higher reaction velocity than at the outer surface. Fine and coarse starch fractions differ not only in total granular surface but also in diameter and volume of their pores. The fine fractions of potato starch, wheat starch and cornstarch have the largest specific surface, pore volume and pore diameter. The relation between pore characteristics and gelatinization has been demonstrated (19).

Huber and BeMiller contest these conclusions (16). They point to the fact that the results are dependant on the method of preparation used. They also claim (14) that potato starch granules do not posses pores and also in all probability no channels. The interpretation in the previous of the term "microchannels" is therefore not clear.

Using non contact atomic force microscopy it was possible to study the surface smoothness of potato starch and tapioca starch granules. In both cases the surfaces are not completely smooth, there are regions with peaks and valleys besides regions which are completely smooth. Tapioca starch granules are smoother than potato starch granules. Also in potato starch, pores are presumed present (20).

Changing the surface, active surface spots

The dense surface layer of the granule also plays a role during physical modification. Balwin describes (21) the effect of ball-milling on the hydration of the "damaged" granules and relates this to the destruction of the protecting surface layer. Besides resulting in a higher hydration, enzymes also show a greater effect. Exhaustive ball-milling finally results in amorphous starch (22).

Heat-moisture treatment (110 °C, 30 min, 20% moisture) changes the X-ray spectrum from B-type to A-type and as can be seen from SEM to changes in the morphology of the granule. A slight damaging of the granule surface results and a void is formed in the center while the surface layer is strengthened (23). This could be the explanation for the retardation in swelling of heat-moisture treated starch. It is also noted that heat-moisture treatment of normal potato starch retards the gelatinization more than a similar treatment of amylopectin potato starch.

Apparently amorphous amylose seems to play a role during this treatment (24). The influence of heat-moisture treatment on a subsequent derivatization is virtually unknown. The same is true for the opposite, derivatization followed by a heat-moisture treatment. An example is the acid hydrolysis of non gelatinized starch granules such as potato starch and high amylose corn starch followed by a heat-moisture treatment. This results in a boiling stable granular resistant starch, important in food applications (25).

It is assumed that the effect of chemical cross-linking is related to the action of the reagent on the outer layer of the granule (13). It has been shown that cross-linking of granular sorghum starch with $POCl_3$ under alkaline conditions takes place predominantly at the surface. This was proven with electron microscopy of granules with thallium as counter-ions of the phosphate groups (16).

Modification of the surface structure with enzymes prior to chemical derivatization hasn't been studied yet. This presents an interesting domain with exceptional surfaces like valleys and craters. There are interesting possibilities which merit further research. The reactivity of starch granules for hydroxyalkylation was enhanced by prior treatment of the starch granules with α-amylase and glucosidases at 40 °C - 45 °C (26). This resulted in higher substitution in alkaline suspension without swelling.

Glucoamylases do have a starch binding domain separated from the active sites. In a 1-3 % glucoamylase solution at pH 3.5 - 6 at 40 °C to 55 °C and a short exposure 0.1 - 15 min. the enzyme is bound to the surface without remarkable hydrolysis. By quickly (5 min.) decreasing the pH to 2.0 the enzyme is deactivated; by increasing the pH to 6.0 the resulting starch shows a hydrophobic surface with a high affinity for oils and fats (27).

Usually α-amylases and glucoamylases are being used in the enzymatic degradation of granular starch. An exception is the degradation with iso-amylases (28). This merits the investigation of the action of other enzymes in the degradation of starch granules.

Surface substitution

Surface substitution occurs during various chemical derivatization in alkaline aqueous slurries. Granule fractionated surface substituted starch derivatives show a higher degree of substitution for fine granule fractions than for coarse granule fractions. When after fractionation of a starch ether or ester into a fine and coarse fraction the fine fraction shows a higher degree of substitution than the coarse fraction, this then indicates surface substitution.

Table 1 shows for a number of starch derivatives the DS of fine and coarse fractions. Starch acetates made with acetic anhydride and starch benzoates made with benzoyl chloride or benzoyl thiosulphate are heterogeneous and indicate probably surface substitution. From these results and other experiments not mentioned here, it can be concluded that surface substitution occurs when using high reactive reagents (fast reactions). Steeneken (29) has extensively investigated this phenomenon.

	DS	d>37μ (%)	D<37μ (%)	D d>37μ	DSd<37μ	DSd<37μ/ DSd>37μ
Acetic anhydride	0.108	46	54	0.085	0.124	1.57
Vinyl acetate	0.081	45	55	0.079	0.076	1.0
Phenylglycidyl ether	0.11	55	45	0.109	0.107	1.0
Benzylchloride	0.15	68	32	0.15	0.15	1.0
Ethylene oxide	0.102	45	55	0.104	0.101	1.0
3-chloro-2-(hydroxypropyl) Trimethylammoniumchloride	0.036	44	56	0.037	0.037	1.0
Benzoylchloride	0.063	49	51	0.040	0.084	2.1
Benzoylthiosulfate	0.079	50	50	0.066	0.095	1.44

*fractions obtained by wet sieving.

Zhenghong Chen et al. (30) investigated the acetyl substitution pattern of acetylated potato- and sweet potato starch as function of the granule size. The DS of the amylose populations of differently sized granule fractions was quite constant while the DS of the amylopectin populations increased with decreasing granule size. The acetylation occurs throughout the amorphous regions and only take place in the outer lamellae of crystalline regions of starch granules.

Surface substitution means that the reaction yield of a small granule starch can be higher than that of a coarse starch like potato starch. In case of regulations imposed by Food and Drug authorities or environmental agencies this can be an important aspect. It is not clear if surface substitution is related to the specific surface layer structure.

During grafting of granular starch often surface polymerization is observed. Recently emulsion polymerization of styrene in the presence of starch was investigated; surface grafting was observed (31). The effects of surface substitution on the physical-chemical properties are hardly known. The difficult gelatinization of potato starch benzoates is probably due to both surface substitution and the dense surface layer and the character of the substituent. Derivatives of benzoyl thiosulphate are easier gelatinized than those of benzoyl chloride (32). The fact that little information can be found about the differences in properties of derivatives with acetic anhydride and vinyl acetate is remarkable in view of the industrial importance of starch acetates.

Hydrolysis

The acid hydrolysis of granular (tapioca) starch in dilute 6% HCl shows the effect of the surface (33). During hydrolysis, the crystallinity increases and the amylose content decreases. After 196 hours hydrolysis at room temperature, a surface layer has been removed and the surface is smooth. The fact that this surface corrosion takes place over the total surface indicates a regular distribution of amylose over the surface. It is well known that heterogeneous acid hydrolysis at room temperature starts within the granule in the amorphous areas (34). This fact is used to determine the distribution of substituents of starch ethers over amorphous and crystalline areas (35).

Ghosts

During gelatinization of starch the resulting aqueous dispersion shows clearly residual granules. These so called "ghosts" are related to the outer layers of the granules that are not completely gelatinized (36). A Japanese patent (37) describes the gelatinization of the outside of a starch granule by a short exposure to steam combined with a microwave treatment which results in completely soluble starches, alpha-starch.

Substituent distribution

The derivatization of alkaline suspended native granular starch with different reagents is of industrial importance. The distribution of the substituents over the mass of starch is virtually unknown. However it has become evident that the crystalline areas are less accessible during the normal slurry reaction. Hood and Mercier (38) already proposed, based on the study of the degradation of hydroxyl propyl di-tapioca starch by α-amylase and pullunanase in 1978 a model where the substituents on amylopectins resided mainly in the non-crystalline areas, near to the branching points. Kavitha and BeMiller (39) concurred for hydroxylpropyl potato starch that the substituents on amylopectins are found near the branching points. A possible explanation for the slight tendency for retrogradation of cross-linked waxy corn hydroxypropyl ethers could be that the substituents are close to the branching points (40).

Richardson et al. (41) studied the substituent distribution of hydroxypropylated potato starch with a combination of high-performance chromatography and electrospray mass spectrometry. She concluded there is a random distribution of the hydroxypropyl groups over the amylopectin chains for granular derivatives (prepared in alkaline slurry). However, in an earlier publication (42) she concluded that for the same products there is a higher degree of substitution near the branching points.

Van der Burgt et al. (43) investigated the distribution of methyl substituents of potato starch. After extensive degradation with α-amylase and precipitation in alcohol two fractions were obtained corresponding respectively with the branched and linear parts of the amylopectin molecule. After determination of the DS it was shown that the methylation of the starch in aqueous suspension with dimethylsulphate occurred preferably at the amorphous parts near the branching points. It also appeared that the DS of amylose was higher than at the linear parts of the amylopectin. Shi and BeMiller (44) demonstrated by fractionating a corn hydroxypropyl ether after aqueous leaching that the amorphous parts of the starch granule are well accessible for hydroxypropylation.

In 2-nitropropyl starch the substituents are almost exclusively bound to the amylose molecule. This could indicate a relationship with initial complex-building between the starch and reagent (45).

The study of enzymatic and acid hydrolysis of cationic waxy corn starch also showed that cationization in aqueous slurry predominantly occurs in the amorphous areas of the granule, especially near the branching points of the amylopectin molecules. A study of dry cationized products concluded that the substituents were mainly attached to the surface and in the channels (46). Similar results were found for cationic potato starches. It was also concluded that besides substitution at the amorphous parts, substitution near the non-reducing ends of the chain also occurred (35).

When oxidizing granular corn starch with hypochlorite (pH 9.5) it was found that the amylose is more accessible for oxidation than amylopectin (47).

A Kansas State University patent (48) describes the cross-linking of partly gelatinized starch granules. To a dispersion of swollen granules with still a crystalline phase a cross-linking agent is added under such conditions that no complete gelatinization occurs. After cross-linking the starch granules they are heated in ample water in order to "melt" the crystalline phase. The material is subsequently spray-dried.

These products retain their individual starch granules after repeated cycles of hot and cold water swelling and drying. After drying the starch granules redisperse immediately in hot or cold water showing complete swollen opaque granules with almost no dissolved material. Application for food, cosmetic and pharmaceutical products is likely. Here the derivatization method used is based on the specific details of the native granular structure.

Drying of starch, never dried starch

Never dried starch or prime starch is starch used without undergoing a drying process.
Never dried starch has hardly been researched in the past.

It has been known for a long time in the starch industry that drying of starch with hot air results in changes in the granule structure which have important effects on the reactivity of the granule. This could be connected to a slow (re)association of the starch molecules (H-bridging). The decrease in reactivity after drying can pose problems when preparing high substituted derivatives.

In this case it is proposed to carry out the substitution in non-aqueous solvents under complete exclusion of water (49). With cellulose, the thermal removal of water leads to less satisfying results in terms of degree of substitution and reaction velocity. The treatment of starch with hydrophilic solvents leads to a complete dewatering of the starch. Dewatering and defatting of corn starch with ethanol results in microporous starch that is used as an aroma binder (50). The dewatering with alcohols is an essential part in the preparation of micro-cellular starch foams from retrogradated starch gels (51).

In the preparation of starch based tabletting agents, it was found that the method of dewatering influences the properties of the end product. Dewatering with ethanol, isopropylalcohol, methanol and acetone led to a higher tensile strength of the tablets than air-, vacuum-, spray or pneumatic drying. Freeze drying had about the same effects as dewatering with alcohol (52). Although the above mentioned cases deal mostly with non-native granular starch derivatives it seems that hot air drying leads to a less open structure in the native starch granule.

Chen (53) states that the drying process increases the formation of compact starch granules which prevents reagents from penetrating into the granules. Prime starch on the other hand, is more open and loose in structure which offers better accessibility for reagents. In addition to these benefits, it costs about two thirds of the cost of dried starch and has the perfect pH range for chemical derivatization (pH 4.0 - 4.5) without additional acid.

When grafting and extruding corn starch, never dried starch (prime starch) has advantages over conventional starch in terms of homogeneity, energy etc. (54). Chen and co-workers (53) studied the grafting of starch with (poly) methylacrylate. Starch graft polymethylacrylate (S-g-PMA) was prepared by Ceric ion initiation of methylacrylate in an aqueous starch slurry (prime starch) which maximized the accessibility of the starch for graft polymerization, no unreacted methylacrylate monomer remained. The prepared grafted starch was then converted to resin pellets and loose-fill foam by single and twin screw extrusion. The use of prime starch significantly improved the physical properties - compression strength and resiliency and water and moisture resistance - of the final loose-film foam in comparison to foam produced from regular dry corn starch.

Zobel (55) mentions in his article *"In my view, the spherulitic, optical and crystallization properties develop as the mature granule dries"*. This conclusion is based largely on having found that the immature starches give amorphous X-ray patterns. Thus granules form and then appear to crystallize. This makes it highly probable that during the industrial drying of starch the crystallinity of the granules changes.

Never dried starch granules are characterized by the presence of substantial smaller cavities than the conventional dried starches. Huber and BeMiller (14) propose a relationship between the presence of pores, channels and cavities and the accessibility of the granule for reagents.

Starch ethers and starch esters prepared from granular starch in alkaline slurry are usually twice dried. One wonders where these products deviate from their "never dried" analogs in physical and application properties.

References

1. G.T.Oostergetel, E.F.J. van Bruggen, Carbohydr. Polym. 1993; 21: 7-12.

2. T.A. Waigh et al., Biopolymers. 1999; 49: 91-105.

3. D.G.Gallant et al., Carbohydr. Polym. 1997; 32: 177-91.

4. H.R.Tang et al., Carbohydr. Polym. 2000; 43: 375-387.

5. Jay-lin Jane et al., Carbohydr.Res.1993; 247: 279-290.

6. D.D.Pan, Jay-lin Jane, Biomacromolecules. 2000; 1: 126-132.

7. K.Koch, Jay-lin Jane, Cereal Chem. 2000: 77(2): 115-120.

8. T.Kasemsuwan, Jay-lin Jane, Cereal Chem. 1994; 71(3): 282-287.

9. N.J.Atkin et al., Starch/Stärke. 1999; 51: 163-172.

10. N.J.Atkin et al., Carbohydr.Polym. 1998; 36: 173.

11. N.J.Atkin et al., Carbohydr.Polym. 1998; 39: 193.

12. T.Galliard, P.Bowler, Starch: Properties and Potential. Wiley. 1987; 55.

13. J.N.BeMiller, Starch/Stärke. 1997; 49: 127.

14. K.C.Huber, J.N.BeMiller, Carbohydr. Polym. 2000; 41: 269-276.

15. K.C.Huber, J.N.BeMiller, Cereal Chem. 1997; 74(5): 537-541.

16. K.C.Huber, J.N.BeMiller, Cereal Chem. 2001; 78(2): 173-180.

17. J.A.Gray, J.N.BeMiller, Cereal Chem. 2001; 78(3): 236-242.

18. Antti Hamune, Starch/Stärke. 1995; 47: 215.

19. T.Fortuna et al., Int. J. Food Sci.Tech. 2000; 35: 285-291.

20. L.Juszczak et al., Starch/Stärke. 2003; 55: 1-7.

21. P.M.Balwin et al., Starch/Stärke. 1995; 47: 247.

22. Y.J.Kim et al., Carbohydr.Polym. 2001; 46(1): 1-6.

23. A.Kawata et al., Starch/Stärke. 1994; 463.

24. K.Svegmark et al., Carbohydr. Polym. 2002; 47: 331-340.

25. WO 00/73352 (7/12/2000).

26. USP 6051700 (18/04/2000) Grain Processing Co.

27. WO 02/24938 (28/03/2002) Grain Processing Co.

28. WO 01/21011 (2001) AVEBE.

29. P.A.M. Steeneken, personal communication.

30. Zhenghong Chen et al., Carbohydr. Polym. 2004; 56(2): 219-226.

31. Chang G. Cho et al., Carbohydr. Polym. 2000; 48: 125-130.

32. NO 165755 (16/04/1981) AVEBE.

33. Napaporn Atichokudomchai et al., Starch/Stärke. 2000; 52: 283-289.

34. R.Hoover et al. Food Rev. Int. 2000; 16(3): 369-392.

35. R.Manelius et al., Carbohydr. Res. 2000; 329: 621-633.

36. R.D.M. Printice et al., Carbohydr. Res. 1992; 227: 121-130.

37. JP 08112067 (7/05/1996).

38. L.F.Hood, C.Mercier, Carbohydr. Res. 1978; 61: 53-66.

39. R.Cavitha, J.N.BeMiller, Carbohydr. Polym. 1998; 37(2): 115-121.

40. Ya-jane Wang et al. Starch/Stärke. 2000; 52: 406-412.

41. S.Richardson et al., J. Chromatography. 2001; 917: 111-121.

42. S.Richardson et al., Carbohydr.Res. 2000; 328: 365-373.

43. Y.E.M. van der Burgt et al., Carbohydr.Res. 1998; 312: 201-208.

44. Xiaochong Shi, J.N. BeMiller, Starch/Stärke. 2002; 54: 16-19.

45. A.Heeres et al., Carbohydr. Res. 2001; 330: 191-204.

46. R.Manelius et al., Cereal Chem. 2000; 77(3): 345-353.

47. Ya-jane Wang et al., Carbohydr. Polym. 2003; 52: 207-217.

48. WO 99/64508 (16/12/1999. Kansas State University.

49. Richter, Augustat, Schierbaum, Ausgewahlte Methoden der Stärkechemie. Wissenschaftliche Verlagsgesellschaft MBH. 1968; 45.

50. A.Smelik, Chechoslovak P 272,971 (1991).

51. USP 5,958,589 (1999) USDA.

52. NOA 9401572 WO 96/09815 AVEBE.

53. L.Chen et al., Biomacromolecules. 2004; 5(1): 238-244.

54. WO 9600746 (1/1/1996) Uni Star Ind.

55. H.F. Zobel, Starch/Stärke. 1988; 40: 44-50.

Derivatization in aqueous alcohols

A: native starch

Most industrial chemical derivatization of starch is carried out in aqueous solution, slurry or under semi-dry conditions (the use of extrusion is a good example). Derivatization in organic solvents is industrially not widely practiced.

The reaction temperature, alkali concentration and DS are limitations in the derivatization of starch in aqueous alkaline slurry. Care has to be taken to avoid swelling of the granules. This makes filtration and washing with water impossible. Mostly swelling inhibitors are used to increase the span of the reaction possibilities. As large amounts of NaCl or Na_2SO_4 are used for this purpose this practice introduces large environmental problems. Reactions in aqueous alkaline solution and isolation of the reaction products with drum-drying introduces salt in the end product making them unfit for food use. Using an extruder poses the same problem.

Washing the end product with alcohol might be the solution, but is hardly practiced because of the cost of the alcohol. Recovery of the alcohol is necessary, but the technology for the recovery of solvents is not widely spread within the starch industry.

The modification of starch in water/alcohol slurry has the following advantages over the modification of starch in aqueous slurry:
• high yields;
• extreme reaction conditions are possible;
• high DS;
• less salt used, so less environmental load;
• different distribution of substituents;
• different product properties.

In this context organic solvents can be divided into three categories:
• solvents in which starch dissolves (DMSO, DMF, pyridine etc.);
• solvents in which starch does not dissolve but which are miscible with water, (alcohol, acetone);

- solvents in which starch does not dissolve and which are not miscible with water (CCl_4, toluene etc);

Although in all these solvents starch reactions can be carried out generally only water miscible solvents - alcohol and isopropylalcohol - are industrially used.

Carboxymethylation

A good example of the derivatization of starch in aqueous alcohol is the carboxymethylation of starch for the production of tablet disintegrating agents.

Thijssen et al. (1) studied the carboxymethylation of granular native potato starch in various solvents (methyl-, ethyl-, isopropylalcohol, three isomeric butylalcohols and acetone). The highest yield was obtained in isopropylalcohol containing 10% of water. The reaction temperature hardly influenced the yield. Also the ratio NaOH/monochloroacetate was studied.

To the starch slurry NaOH was added as pellets and the mixture was stirred overnight at the reaction temperature to ensure equilibrium between the hydroxide and the starch. Sodium monochloroacetate was added in dry form. When carrying out this reaction as a one stage reaction, a DS up to 1.3 was reached with a reaction efficiency of 60%. After carrying out the reaction in three subsequent steps, a DS of maximum 2.2 was reached.

A similar procedure for carboxymethylation is patented by Wolff Walsrode (2). The presence of small amounts of water in the reaction mixture is essential to realize a high DS with a high yield and to prevent the build up of lumps. If the DS is sufficiently high the products dissolve completely as a clear solution.

Heinze and co-workers (3) studied the carboxymethylation of starch in a methanol/water mixture. When the reaction is carried out in one step the DS is less than 0.6. There is mainly monosubstitution of the glucose units as was shown by HPLC/acid hydrolysis and ^{1}H-NMR-spectroscopy. In a multi step reaction a DS up to 2.1 can be reached. The reactivity of the OH-group decreases in the following order O-2> O-6>O-3. After complete hydrolysis the composition of the monomer mixture was in good agreement with the classical statistical model of Spurlin (4).

Cationization

Kweon (5, 6) has presented a good survey of the cationization of starch in aqueous alcohol. The low substituted cationic starches are produced with a good yield. The cationization of corn starch with 2-chloro-3-hydroxypropyltri-methylammoniumchloride has an optimum in 65% ethanol at 50 °C and a starch/water ratio of 65%.

Hydroxyethylation

Grain Processing Company patented (7) a method for hydroxyethylation in aqueous alcohol with high DS and sufficient solubility without a gelatinization peak. Although this method has been known for a long time there still is ample room for patentable new special products.

Esterification

As far as known by the authors, industrial esterification is not practiced in aqueous alcohol. High substituted succinic half esters are a possibility.

Cross-linking

Since the 1950's the cross-linking of starch - also in combination with etherification and esterification - with multi-functional reagents like $POCl_3$, epichlorohydrin and sodium tri metaphosphate in dilute alkaline slurry is well known. It has proven to be a highly successful method to produce thickening agents for all kinds of applications. Small amounts of reagent do have large effects as do the reaction conditions e.g. temperature (8). The cross-linking in aqueous alcohol is hardly studied.

Hydrophobic starch

The reaction of granular starch slurry with reagents such as benzylchloride and phenylglycidylether in an aqueous hydroxide solution has been studied intensively during the last years. Low substituted derivatives can be produced with a good yield. However gelatinization of these derivatives in water is difficult. Benzylated derivatives with a DS of 0.2 already have to be dispersed mechanically (9).

This difficult gelatinization of hydrophobic starches is caused by annealing, hydrophobicity of the substituent and heterogeneity. Annealing under reaction conditions causes an increase in the crystallinity and hampers the gelatinization (physical cross-linking). The introduction of a hydrophobic substituent decreases the hydrophilicity of the starch. The negative effect of the hydrophobic group on the solubility of the derivative increases with increasing temperature and is strengthened by the fact that the substituents often are present in clusters, an entropic driven hydrophobic clustering. Etherification and esterification of granular starch in aqueous alkaline slurry predominantly occurs in the amorphous areas. Surface substitution will increase the heterogeneity even further.

Benzylation of starch in aqueous alcohol could be part of the solution for the above mentioned difficulties with gelatinization of these derivatives. Christian et al. (10) demonstrated that during esterification of sucrose with alkylisocyanates in aqueous alcohol less polysubstitution occurs than in water.

Hydrolysis

Hoover et al. (11) has presented a well documented overview of the acid hydrolysis of granular starch in aqueous alcohol. Also in the patent literature the acid hydrolysis of starch in aqueous alcohol is described. By using methanol a thin boiling starch with low alkali number and good film-forming properties is obtained (12). An other patent (13) describes the preparation of a gum arabic replacer with clear film forming properties. Salts are first removed from starch ethers and esters by washing with alcohol. Heating to 150 °C in aqueous alcohol with acid present renders the product cold soluble. At the same time some hydrolysis is observed.

Also the methanolysis of polysaccharides is described (14).

The hydrolysis of non-starch polysaccharides could be of interest. Polyglucuronic acid hydrolyses faster in a water/alcohol mixture than in water.

Fox and Robyt (15) describe the formation of new kinds of limit-dextrins by hydrolysis of granular starch with HCl in various alcohols. By varying the hydrolyzation conditions limit-dextrins with different DP were obtained. Also, they have described the influence of mixtures of alcohols (16). The demonstrated influence of the reaction conditions (acid, solvent) implicates that the glucosidic bonds in the granule have different reactivity.

The method to produce "sequesters" - hydrolysates of by preference amylopectin starches having complexing properties comparable to cyclodextrins - is related to the previous. In the first step the starch is heated in a mixture solvent/non solvent, preferably water/ethanol. In this step the starch matrix is converted to a single helix form. In the second step acid is added (H_2SO_4) and the mixture is heated. In this step specifically the A- and B-chains are removed. These chains act as sequesters (17).

Jane (18) describes the combination of acid hydrolysis with ball milling. Small starch particles are obtained in this way.

The degradation of granular starch in aqueous alcohol with mineral acids is claimed to produce a homogeneous molecular weight distribution. This is explained by the faster degradation of the amylopectin molecules over the amylose. After introduction of polyalkyleneoxide- and sulfate groups these products are used as plasticizer in cement (19).

Oxidation

The oxidation of starch in aqueous alcohol seems at first sight not obvious as alcohol is also oxidized. The oxidation of hydroxypropyl starches with perborate has been described (20). Apparently there is enough difference in the reactivity of the -OH groups of starch and alcohol. Further research is necessary.

Physical modification

Cold water soluble granular starches (CWSG) are prepared by heating starch (150°C - 350°C) during a short time in aqueous alcohol. This preparation makes use of the swelling inhibition of the starch by alcohol. As alcohol, ethanol and iso-propylalcohol are used. The starches based on corn and wheat still show some birefringence, but the characteristic Maltese cross is missing (21). X-ray diffraction shows a V-spectrum. The amylopectin contributes to the spectrum, the V -single helix structure correlates with cold solubility. CWSG starches are used over drum dried starches in food applications with high requirements regarding structure, texture and rheology (pudding). Also the preparation in alcohol/water/alkali is described (22, 23, 24).

A treatment of starch in aqueous propanol below 100 °C mimics a heat-moisture treatment. The products are partly decrystallized and do not swell in cold water. Heat-moisture treated starches - produced by heating starch with a small amount of water - are sometimes applied instead of cross-linked starches. From an environmental point of view this treatment has to be preferred over cross-linking. Amylopectin starch treated in this way behaves differently from 'normal' starch. Apparently the amylose plays a role in this treatment.

In the 1900's the preparation of cold soluble starches was carried out by treating starch in an aqueous alcoholic hydroxide solution (25). A high activated starch is prepared by precipitating an alkaline starch solution in methanol. After air drying the product is used for the preparation of high DS starch acetates (26).

Defatting and drying of corn starch with ethanol increases the carrying capacity of the starch for flavors and fragrances (27).

Enzymatic conversion

Enzymatic conversion of starch in aqueous alcohol is gaining more attention. The degradation of wheat starch in aqueous alcohol (30%) with α-amylases results in oligosaccharides with a high DP (28). A similar conversion with glucoamylase is also known.

Derivatization in aqueous alcohols

B: Pre-cooked starches

In the previous part the derivatization of native starches in aqueous alcohol is discussed. The derivatization of cold water soluble granular starches in aqueous alcohol suspension is hardly industrially used.

The ratio water /alcohol can be varied. Care has to be taken to prevent swelling and dissolution of the starch. In Figure 3 the influence of the water/alcohol ratio on the temperature at which swelling of the drum dried starch starts is presented (Brabender 10% ds). To increase the reaction velocity the reaction temperature has to be increased to temperatures over 40 °C causing the use of high alcohol concentrations.

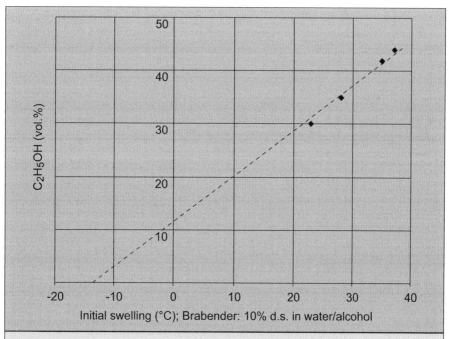

Figure 3. Temperature of initial swelling of a drum dried potato starch in aqueous alcohol.

Preparation of cold water soluble starch slurries in aqueous alcohol

Slurries of cold water soluble starches in aqueous alcohol are prepared by mixing drum dried, extruded or spray dried starches into aqueous alcohol.

In the laboratory native starch is flocculated from an aqueous solution in alcohol. A fibrous or flock-like cold water soluble product results.

Probably this procedure is also possible on a large scale. The flocculation of i.e. alginates in $CaCl_2$/isopropylalcohol with specific equipment for the injection is patented (29). Also the use of a rotor-stator to flocculate native starch in a saline solution is patented (30). It seems obvious to combine the mentioned flocculation procedure with the modification of cold water soluble starches. This makes an intermediate drying step, with possible negative effects, on the reactivity superfluous.

Chemical derivatization

One of the few examples from literature of the chemical derivatization of cold soluble starches in aqueous alcohol is the cross-linking of these starches or mixtures of starches in alcoholic slurry. In alcoholic slurry cross-linked cold soluble starch or mixtures of cold soluble starches are used for the purification of α-amylases (31). From an application point of view there are more possibilities. Thickening agents and super-absorbents could be produced showing improved properties. Also the hydrophobation of starch could be improved caused by a more homogeneous substitution and the amorphous structure, the water solubility or dispersability increases, may be also at room temperature. A combination of different types of derivatization is also an option.

Physical modification

Besides chemical derivatization, physical modification of cold water soluble starches in aqueous alcohol also seems promising. What happens during the heating of cold water soluble starch in aqueous alcohol? Will there be a recrystallization under favorable heat and moisture conditions? It is known that cold water soluble starch in an aqueous saline solution rapidly converts into a hot water soluble product with properties different from native starch.

From Figure 3 can be seen that by extrapolation to an alcohol concentration of 0 a kind of virtual temperature at which swelling starts in water can be determined (-13.5°C). Perhaps it is possible to characterize the different types of cold soluble starches in this way. It might even be possible that if the crystallization of water (freezing) can be avoided, drum dried starch could be modified in water at low temperature without swelling. This is certainly true in the presence of small amounts of alcohol.

Reactions at low temperature

Starch derivatization mostly has been studied at high reaction temperatures. Low reaction temperatures offer interesting opportunities. Reagents which are too reactive under the currently used conditions such as aliphatic acid-chlorides become possible. The distribution of the substituent e.g the homogeneity of the substitution is more favorable.

References

1. C.J. Thijssen et al., Carbohydr. Polym. 2001; 45: 219-226.
2. WO 02/04525 (17/1/2002) Wolff Walsrode.
3. T. Heinze et al., J.Appl. Polym. Sci. 2001; 81(8): 2036-2044.
4. W.Lazik et al., J.Appl. Polym. Sci. 2002; 86: 743-752.
5. M.R.Kweon et al., Starch/Stärke. 1996; 48: 214.
6. M.R.Kweon et al., Starch/Stärke. 1997; 48: 59.
7. WO 99/67295 (1999) Grain Processing Co.
8. K.F.Gotlieb et al., Die Stärke.1967; 19: 263.
9. J.Bohrisch et al., Starch/Stärke. 2004; 56: 322-330.
10. D.Christian et al., Tetrahedron Letters. 2004; 45: 583-586.
11. R.Hoover et al., Food Rev. Int. 2000; 16(3): 369-392.
12. USP 3,193,409 (1965).
13. USP 4,837,314 (1989).
14. E.G.Roberts et al., Carbohydr. Res. 1987; 168(1): 103-109.
15. J.D.Fox, J.F.Robyt, Carbohydr. Res. 1992; 227: 163-170.
16. J.F.Robyt et al., Carbohydr. Res. 1996; 281: 203-218.

17. WO 94/17676 (1994).

18. Jay-lin Jane, Cereal Chem. 1992; 69(3): 280-283.

19. WO 68272 (2000).

20. Eur.P. 067415 (1995).

21. Jay-lin Jane, Starch /Stärke. 1986; 38: 258.

22. J.Chen et al., Cereal Chem. 1994; 71(6): 618-622.

23. J.Chen et al., Cereal Chem. 1994; 71(6): 623-626.

24. L.A.Bello-Perez et al., Starch/ Stärke. 2000; 52: 154-159.

25. USP 4,992,539 (1991).

26. M.Samec, Kolloidchemie der Stärke. 1927; 182.

27. Eur. P. 579,179 (1994).

28. A.Smelik, Czechoslovak P 272,971 (1991).

29. JP 06121693 (1994).

30. Fr.P. 270103 (1994).

31. Ger.Offen DE 4 218 667 (1993).

32. NOA 8901576.

Application of granule swelling inhibitors in derivatization

When preparing starch ethers and esters in aqueous suspension, alkaline catalysts like Na_2CO_3, NaOH and $Ca(OH)_2$ in aqueous solution are used. As acid is formed during the preparation of starch acetates with acetic anhydride, diluted hydroxide is added during the reaction to keep the pH on the desired level of 8.5. The reaction time at room temperature is at the usual low degrees of substitution relatively short, about 30 minutes. With etherification reactions like hydroxyethylation with ethyleneoxide no hydroxide is consumed, however, the reaction times are longer, the pH and temperature are higher.

During these reactions always the possibility exists, especially at high degrees of substitution, that the starch granules will swell under influence of the hydroxide. This is the reason for limitations in reaction conditions and degree of substitution. The limitations differ for the different types of starch, corn starch allows for more extreme conditions than potato starch.

Swelling inhibitors, inorganic salts

The swelling inhibiting effects of adding salts like NaCl or Na_2SO_4 to the reacting starch slurry was discovered rather early. These salts, here called swelling inhibitors, not only retard the swelling of the starch granule but also increase the absorption of hydroxide by the granule. To maintain the same pH in a slurry with swelling inhibitors more hydroxide has to be added than to a slurry without swelling inhibitors. For instance, in a 38% wheat starch slurry, the uptake of hydroxide to maintain a pH of 12 increases with 68%, 88% and 119% after the addition of respectively 4%, 8% and 16% Na_2SO_4 calculated on the dry starch (1).

The effects differ for each type of starch, but also depend on the reaction conditions and DS. Sodium sulfate is more effective than sodium chloride, also Na_2SO_4 gives better results at high concentrations than NaCl. Increasing the amount of NaCl used to over 50 g per kg starch hardly has an effect, while levels of 200g Na_2SO_4 are very common. For the preparation of highly substituted derivatives Na_2SO_4 is advantageously used. Generally swelling inhibitors are only used when starch is etherified, for example, with halogenides and epoxides. The esterification of starch with acetic anhydride can be carried out without a swelling inhibitor.

To prevent swelling during etherification and esterification the concentration of the diluted hydroxide added to the starch suspension has to be below a certain level. For potato starch a 4.5% sodium hydroxide seems to be the maximum. Sometimes a mixture of hydroxide and swelling inhibitor is added reducing the swelling of the starch (derivative) during the addition of hydroxide. At the spots where the drops of the relatively concentrated hydroxide have made contact with the starch suspension swelling starts easily, therefore the reaction mixture has to be stirred vigorously.

In the laboratory this poses no problems, however, under industrial circumstances special measures have to be taken - especially if during the reaction hydroxide has to be added - to ensure a good catalyst action and prevention of swelling.

There is not much literature about the absorption of hydroxide under influence of salts. Some classical publications by Leach, de Willigen, de Groot and Oosten (2, 3, 4) are worth mentioning. More recent is the work of Paul A. Seib and coworkers (1, 5). They found when 4% sulfate was present in a pH 12 medium, hydroxide or alkoxide ion in the starch doubled and 93% of the sulfate ion was present in the starch phase as determined by sulfate assay. In addition, the starch granules assumed a net negative charge and the continuous phase became positively charged. The surprisingly high absorption of sulfate ions inside the negatively starch granule was confirmed by a separate sulfate assay on the supernatant (1). Earlier explanations with regard to hydroxide absorption in the presence of salts such as the one from van Oosten - absorption of sodium ions, Donnan potential effects - needs revision as mainly negatively charged ions are absorbed.

The application of swelling inhibitors makes robust reaction conditions possible resulting in high yields, high DS, shorter reaction times and less use of reagents. Kyungsoo Woo et al. describe a cross-linking reaction of wheat starch with sodium trimetaphosphate in the presence of sodium sulfate (5). The degree of cross-linking was higher.

The following explanation was given. Swelling inhibitors:

- influence the structure of water, enhancing the penetration of the reagent into the granule;
- increase ion strength and promote the reaction of starch alkoxide with ionic phosphorus;
- increase the ionization of starch in the presence of salt;
- increase of the quantity of sodium ions in the granule.

About the same explanation can be given for the effects of swelling inhibitors during the hydroxyalkylation of starch.

Side effects

In addition to the mentioned positive effects of the use of swelling inhibitors sometimes desired and/or undesired side effects do occur. In concentrated alkaline salt solutions at elevated temperatures annealing effects are observed hindering the gelatinization of the produced derivatives.

Sometimes an unwanted "cross linking effect" can be positive. The introduction prior to chemical derivatization of physical cross-links reduces the swelling during the reaction, resulting in a higher DS (6). Use of reversible chemical cross-links prepared for instance with divinylsulfon, in the etherification in alkaline suspension makes, high DS possible. After the reaction the cross-links are removed (7). This reaction is not used on a large scale.

The reaction of reagents with swelling inhibitors can pose problems in later use of the products. During the etherification of starch with propyleneoxide in the presence of sodium chloride, propylenechlorohydrin is produced. This prevents the use of this product in food applications.

After the hydroxypropylation of wheat starch in alkaline slurry the majority of the lysophospholipids are removed in the following finishing steps (1).

Organic swelling inhibitors

The use of swelling inhibitors is becoming more and more an environmental problem. Since recovery of inorganic salts from dilute solutions is expensive the use of organic swelling inhibitors was investigated. Removal by anaerobic digestion looks promising. Villvock (8) found that salts of citric acid are effective in protecting starch granules against swelling. The yield of hydroxypropylation was lower than in the conventional method. Shi studied the differences in the effects of citrate and sulfate during the hydroxypropylation of corn starch and found a preference for substitution of amylose in the presence of citrate (9).

Alcohol generally has not been regarded as a swelling inhibitor although the preparation of derivatives in dilute aqueous alcoholic solutions is well known. The step from inorganic salts to citrate is not larger than from citrate to alcohol.

In the previous chapter the study of Gray and BeMiller (10) about the penetration of fluorescent fatty-amides into the starch granule was mentioned. Two conclusions will be repeated here:
- Swelling inhibitors reduce the penetration of the studied reagents into the granule.
- Sodium citrate is more effective than sodium sulfate.

Finally, the use of enzymes (hydrolases) like α-amylases, amyloglucosidases and isoamylases as a pretreatment of the starch in a slurry prior to derivatization to increase the reactivity (11) has to be mentioned. If this pretreatment reduces the amount of swelling inhibitor is not completely clear.

References

1. Naoko Matsunaga, Paul Seib, Cereal Chem. 1997; 74(6): 851.
2. H.W. Leach, T.J. Schoch, E.F. Chessman, Die Stärke. 1961; 13: 200-203.
3. A.H. de Willegen, P.W.de Grooth, Die Stärke. 1971; 23: 37-42.
4. B.J. Oosten, Starch/Stärke. 1982; 34: 233-239.
5. Kyungsoo Woo, Paul Seib, Carbohydr. Polym. 1997; 33: 263-271.
6. USP 3,577,407 (1971).
7. USP 3,438,913 (1969).
8. V.K. Villvock, M.S. Thesis. Purdue University. 1996.
9. X. Shi, J.N. BeMiller, Carbohydr. Polym. 2000 43(4): 333.
10. J.A. Gray, J.N. BeMiller, Cereal Chem. 2001; 78(3): 236-242.
11. USP 6,051,700 (2000) Grain Processing Co.

Amylose inclusion complexes

There is ample literature about the complexing of various guest molecules by amylose. Amylose molecules can form a helix with the hydroxyl groups of the glucose units arranged on the surface of the helix, the inside of the helix being more hydrophobic than the outside. Well known complexes of amylose are those with butanol and iodine. The iodine complex has a deep blue color and is used in numerous analytical applications. The complexing of butanol became known by the work of Schoch (1, 2) who used the insolubility of the amylose complex in water to separate the amylose from the amylopectin.

Shinkai and co-workers (3) used complexing in preparing "memory polymers" from starch as alternatives for cyclodextrins. Starch was cross-linked in aqueous solution with cyanuric chloride in the presence of a template. When using methylene blue with cyanuric chloride dissolved in benzene, a gel was formed in the water phase. This gel was divided into small parts and extracted with 50 % aqueous pyridine in a Soxhlet until the color disappeared. After washing with methanol the product was vacuum dried. After swelling in water the resin (50 mg in 10 ml water) absorbed methylene blue quantitatively at 30 °C. It was calculated that 9.5 glucose units are involved with the bonding of the template. For a resin produced in the same way but without a template 59 glucose units were involved with a bonding constant 2.4 times smaller. Shinkai states that taylor made "cyclodextrins" prepared in this way are applicable more precisely and successfully to many novel phenomena attained in the past by use of cyclodextrins.

To avoid problems caused by the insolubility of the amylose complexes Wulff and Kubic (4) used a hydroxypropyl amylose (DS 0.075). This made it possible to measure Cotton effects. Amylose and α-CD complexes with 2-hexanon gave a negative Cotton effect. β- and γ-CD give positive Cotton effects. In the first case 6 glucose units per helix are involved. Using 4-t-butylphenol amylose and α-CD gave different Cotton effects. Apparently amylose can complex the more bulky molecule while α-CD cannot. β- and γ-CD now show the same Cotton effect as amylose and 7 glucose units per helix are involved. Phenolphtalein is complexed at the ends of the amylose molecule. The stability of the complexes depends on the chain length of the amylose molecules.

Based on microcalorimetry experiments it was concluded that sodium dodecyl sulfate (SDS) was complexed at a DP of 9. At higher DP values the heat effects increase till DP 200. Amylose with a DP <250 did not complex 4-t-butylphenol (TBF). The complexing of SDS is faster than TBF. This is explained by the amylose having already 6 glucose unit helixes where the SDS molecule can fit in easily. With TBF a conformation change is first necessary and this is a slow process even impossible at a DP < 250. The authors also discuss the effects of cross-linking. Cross-linking at a low concentration with epichlorohydrin (2%) resulted in intra molecular cross-linking, based on the measured molecular weights. Iodine absorption decreases at increasing cross-linking. It is assumed that amylose under these circumstances has a random coil conformation making iodine complexes impossible. Cross-linking freezes this conformation; hence the cross-linked products also do not complex iodine. When cross-linking was carried out in the presence of cyclohexanon the iodine absorbtion increased. Apparently the presence of the complexing agent promotes a more fixed helix structure.

For intermolecular cross-linking amylose complexes were cross-linked with cyanuric chloride. The resulting products could be isolated as insoluble resins. After washing with water and methanol the products were amylose and complex builder free. Intermolecular cross-linked products can give complexes with various molecules depending on the character of the host molecule during the cross-linking.

Cyclodextrins can also complex high molecular weight molecules and polymers, α-CD gives a complex with polyethylene glycol if the M_w is > 300, β-CD and γ-CD cannot form these complexes; β-CD gives exclusively complexes with polypropyleneglycol and γ-CD with polymethylvinylether (5, 6). When several molecules CD contain one molecule of polyethyleneglycol they are called polyrotaxanes. The cross-linking of PEG -rotaxanes with epichlorohydrin in alkaline conditions under the formation of diglycerylethers is also described (Figure 4). Such rotaxanes are closed at the outer ends with bulky groups. After removal of the bulky group and the complexed polymer so called polymer tubes are left. Comparison of such polymer tubes with cross-linked amylose might give more insight in the properties of both products. Kadokawa (7) complexed poly-tetrahydrofuran with amylose by polymerizing glucose-1-phosphate with fosforylase to amylose in the presence of poly-THF.

Gruber and Konisch describe (8) the increase in viscosity of aqueous cellulose derivatives with hydrophobic substituents by adding amylose to the solution. Hydroxyethyl cellulose was converted into a cationic derivative with dodecyl groups attached to the quaternary nitrogen. This derivative was dissolved in

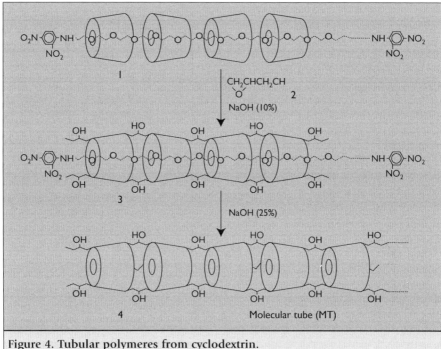

Figure 4. Tubular polymeres from cyclodextrin.

the presence of amylose and showed a higher viscosity than the non hydrophobic derivative. In the presence of CD's and amylopectin this effect is not observed. An explanation could be the complexing of the hydrophobic groups by the amylose. (see Figure 5). The resulted cross-linked "clathrate network" is

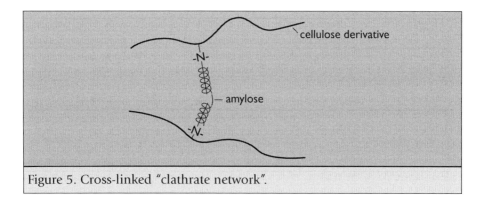

Figure 5. Cross-linked "clathrate network".

pseudoplastic in aqueous solution and temperature stable. Heating of the complex reduces the viscosity which returns slowly upon cooling.

Aqueous solutions of polyvinylalcohol and starch are instable; upon standing phase separation will occur. As such systems are important for the paper- and textile industry starch derivatization was applied to improve the miscibility of starch with pva. Degraded hydroxyethyl starch ethers are better miscible with pva. Okaya and coworkers found (9) that derivatization of pva improved the miscibility with starch solutions. End group modified pva (octadecyl) showed specific interaction with starch thereby increasing the miscibility with oxidized starch. PVA modified with long alkyl groups as side-chains showed the same behaviour. Model experiments with amylose and amylopectin showed complexing as an explanation, with amylopectin more specific than with amylose.

National Starch describes in a patent (10) the application of the complexing with amylose. Amylose containing starch is converted with maleic anhydride derivatives into derivatives containing substituents with at least 12 linear hydrocarbons. In the preparation a phase-transfer-catalyst is used. Gelatinized solutions of such derivatives form gels at temperatures of about 70 °C or higher - dependant on the type of starch, DS, type of substituent. This gelling is explained by the complexing of the hydrophobic substituents by amylose.

The oxidation of starch ("amylose dextrins") and cyclodextrins in the presence of complex builders like alkane carbonic acids and alkane sulfonic acids with 3 to 19 carbon atoms or the equivalent alcohols with 4 to 16 carbon atoms is described. With glucans and similar polysaccharides the oxidation is specifically at the C_6 hydroxymethyl group (11). Oxidation with the conventional oxidation agents is possible, but the electrochemical oxidation with metal-electrodes is preferred. There is no unwanted reaction to a ketone, no ring opening to a dialdehyde or dicarbonic acid and no depolymerization. This is explained by the formation of a complex and helix structure making the C_6 hydroxyl groups better accessible. The specific C_6 OH oxidation was indicated by ^{13}C NMR.

Heeres (12) converted starch in aqueous alkaline slurry with 2-nitropropylacetate.

nitropropylacetate

After aqueous leaching the amylose derivative could be separated from the amylopectin derivative. The partition of the substituents over the two fractions is given in Table 2.

Table 2. Distribution of substituents over amylose and amylopectin fractions.

Derivative	Reagent	DS(amylose)	DS(amylopectin)	DS ratio
Methyl	$(CH_3)_2SO4$	0.147	0.095	1.5
2-nitropropyl	2-npa	0.554	0.032	15
2-nitropropyl	2-npa	0.772	0.015	50

In 2-nitropropylstarch the substituent is almost exclusively bound to amylose. This is probably explained by the complexing of the 2-nitropropylacetate by the amylose after which the C-OH groups reacts.

References

1. T.J. Schoch, Cereal Chem. 1941; 18: 121.
2. T.J. Schoch, J. Amer. Chem. Soc. 1942; 64: 2957.
3. Seiji Shinkai et al., Tetrahedron Letters. 1983; 24(33): 3501-3504.
4. W. Kubick, Starch /Stärke. 1993; 45: 220.
5. A. Harada et al., Nature. 1992; 356: 325.
6. A. Harada et al., Macromol. Rep. 1995; A32(supl5,6): 813-819.
7. Jun-ichi Kadokawa et al., Chem. Commun. 2001; 449-450.
8. J.V. Gruber et al., Macromolecules. 1997; 30: 5361-5366.
9. Takuji Okaya et al., J. Appl. Pol. Sci. 1992; 45: 1127-1134.
10. EP 0188237 (1986) National Starch.

11. NOA 8902428 (1991).

12. A. Heeres, Thesis. University of Groningen. 1998.

Starch derivatization – Fascinating and unique industrial opportunities

Synergistic effects

The viscosity of mixtures of dilute (1.2%) aqueous solutions of hydroxyethyl cellulose (HEC) and carboxymethyl cellulose (CMC) was measured at different shear and temperature by Zhang (1). All mixtures showed viscosity synergy with a maximum at 67% HEC. The products used were obtained commercially from Hoechst and used as such. Based on the UV- and IR-spectra it was concluded that the synergy probably is connected with hydrogen bridges between the carboxyl- and hydroxyethyl groups. However to appreciate the observed effects it is necessary to know more about the molecular weights, salt content of the products and the pH of the aqueous solutions and in particular the ratio between the -COOH- and -COO-groups. Also cross-links caused by internal esterification of the -COOH groups can play a role.

The starch industry uses hydroxyethyl cellulose already for a very long time to 'upgrade' (cross-linked)carboxymethyl starches used as thickener in technical applications. If synergy in this case plays a role in this application is not clear.

Zhang claims in his publication to be the first to deal with properties of aqueous solutions of mixtures of HEC and CMC. This is a bit strange as in old patent literature (2, 3) the properties of cellulose mixed ethers are already mentioned. Solutions of hydroxyethyl/carboxymethyl cellulose ethers show high viscosities especially in aqueous salt solutions.

In some applications of carboxymethyl starch (CMA) the sensitivity of the viscosity for Ca ions is important. This is observed in the application of CMA in poster-pastes with $CaCl_2$ added as anti-freeze. In the application of these products in super absorbents and drilling mud additives the sensitivity for NaCl is observed. In the application of CMA as thickener, these products are mostly not used as such but after cross-linking. Cross-linking with epichlorohydrin results in higher (structure) viscosity and improvement against shear and pH. The viscosity of some cold water swelling carboxymethyl starches with different degree of cross-linking and hydroxyethylation is presented in Figure 6.

The CMA in aqueous $CaCl_2$ (within the indicated concentrations) hardly develops a viscosity. Cross-linking also shows hardly an effect. Additional hydroxyethylation shows both in aqueous as in aqueous $CaCl_2$ solutions a much higher viscosity. This viscosity increase of the product in pure water can be

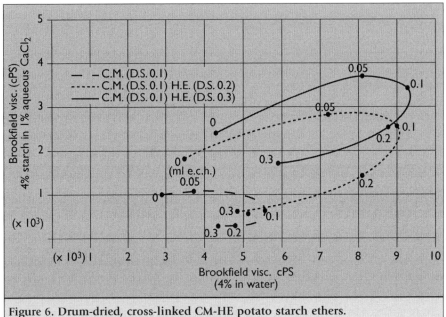

Figure 6. Drum-dried, cross-linked CM-HE potato starch ethers.

explained by the salt content of the products (introduced during the production) causing salt effects. Increase of the DS increases the viscosity further although the effect levels off. Subsequent cross-linking increases remarkably the viscosity. The introduction of both carboxymethyl- and hydroxyethyl groups leads to a compromise between a high viscosity in water and a high viscosity in an aqueous salt solution.

This leads to the question if mixed ethers do have a better performance than mixtures. So far this question - not an easy one as all products to be investigated have to be made from the same raw materials and the same chemicals under identical conditions - has not been studied.

Another issue - for those cases where viscosity is not predominant - is the use of carboxymethyl/hydroxyethyl ethers of starch, cellulose etc. In the preparation of super absorbents (4, 5) based on CMC and HEC in one case mixtures of CMC and HEC are cross-linked with divinylsulfon while in the other case mixtures of CMC and HEC are used. In applications like sequestrants, drilling mud additives, textile printing pastes etc. there are possibilities for carboxymethyl starches and carboxymethyl celluloses whether cross-linked or not in combination with hydroxyethylation. Also the combination of hydroxyethylation and the introduction of a charged substituent different from the carboxymethylgroup might result in interesting effects. Cationic starches or starch phosphates with additional hydroxyethyl groups are the candidates.

The partial hydroxyethylation of carboxymethyl starch and cellulose with ethyleneoxide is described (6). In this reaction the hydroxyethyl esters of the caboxymethylethers are formed. The aqueous solutions show high viscosities and high electrolyte resistance. This demonstrates the effectivity of a hydroxyethyl group on a spacer. The study of other non-charged substitutes like dihydroxypropyl on the rheology of carboxymethyl derivatives might be important.

Hydroxyethylethers of potato starch can be oxidized with hypochlorite and 2,2,6,6,-tetramethylpiperidinyloxy (TEMPO)/Br- as catalyst into carboxymethyl-hydroxyethyl derivatives of C_6-carboxypotatostarch. Such products can be used as sequestrants (7). Also, here the study into the influence of the hydroxy ethyl groups might be worthwhile.

The synergistic effects in the application of starch derivatives are only incidentally reported.

Some examples:
- the use of 'synergistic' mixtures of cationic potato- and wheat starches in paper making (8);
- the use of mixtures of a sugar or an oxidized sugar and a hydrogenated sugar to increase the plasticity of mineral binders in cement (9);
- the use of a mixture of a linear polymer with a branched starch in an aqueous surfactant solution in household detergents and tooth paste (10);
- the use of mixtures of starch based super absorbents with dextrins to increase the absorption capacity of the polymer (11).

References

1. L.M. Zhang, Colloid Polym. Sci. 1999; 277: 886-890.
2. USP 2,618,632 (1952) Hercules.
3. USP 3,446,795 (1965) Dow.
4. F. Esposito et al., Polym. Mater. Sci. Eng. 1996; 74: 368-369.

5. W. Opperman, Das Papier. 1995; 765.

6. USP 3,092,619 (1963).

7. WO 9638484 (1996) AVEBE.

8. EP 139 597 (1985) Roquette.

9. USP application 20030010254.

10. USP 5,286,405 (1995).

11. USP 4,483,950 (1984).

Carbohydrate oxidation with oxygen, varieties on the Spengler- and Pfannenstiel reactions

Aldoses are stable in neutral aqueous solutions. Under alkaline conditions enolisation, isomerisation and "alkaline degradation" occurs (1). In the presence of oxygen under neutral conditions there occurs no reaction, but under alkaline condition an aldose is converted into the next lower aldonic acid. This reaction is the Spengler- and Pfannenstiel reaction (2). The reaction mechanism is illustrated below.

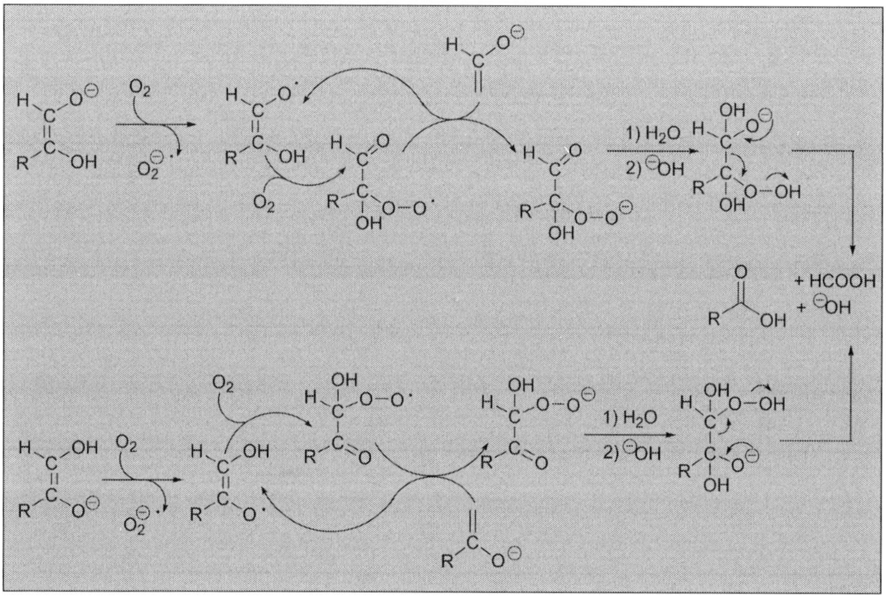

The yield is usually about 60% - 75%. When the oxygen concentration is (too) low the earlier mentioned reactions occur as enolisation etc.

A patent of Cerestar (3) describes the Spengler- and Pfannenstiel reaction with maltodextrins. "In a reactor 400 g. maltodextrin with a DE of 11-14 is dissolved in 0.5 l. demineralized water. The solution is saturated with oxygen and 306 g. of a 45% NaOH solution (w/w) in water is added. The oxygen pressure is kept at 2 bars *and the reaction mixture is heated to 60°C. After 2 hours the reaction is terminated. The pH of the brown colored solution is about 11. With H_2O_2 the solution is bleached and if necessary neutralized with citric acid. The reaction mixture consists of 56% polymeric material and 44% degradation products (CO_2, formic acid, glycolic acid, glycerinic acid and dihydroxybutyric acid)".*

Analysis of the hydrolyzed polymeric material shows that the reducing end groups of the maltodextrin are oxidized into arabinonic acid. In this patent it is also mentioned that increase of the alkali concentration increases the polymeric fraction and addition of Fe^{++} increases degradation. The latter is possibly caused by cleaving of the chain resulting in new end groups leading to the so called peeling reaction (4). The products are proposed for use in the paper making and textile industry and as co-builders in detergents.

A Novamont patent (5) describes the oxidation of starch with oxygen under alkaline conditions in the presence of metal-ions such as Cu, Cr, Co, Ni, Fe, V, Mn and Ti. In another patent Novamont claims the use of complexes such as $FeCl_2$-o-phenanthroline as catalyst (6). These are all Spengler- and Pfannenstiel reactions but now combined with catalysts.

Hendriks (7) describes the Spengler- and Pfannenstiel reaction with lactose and reports a selectivity of 75%-80%.

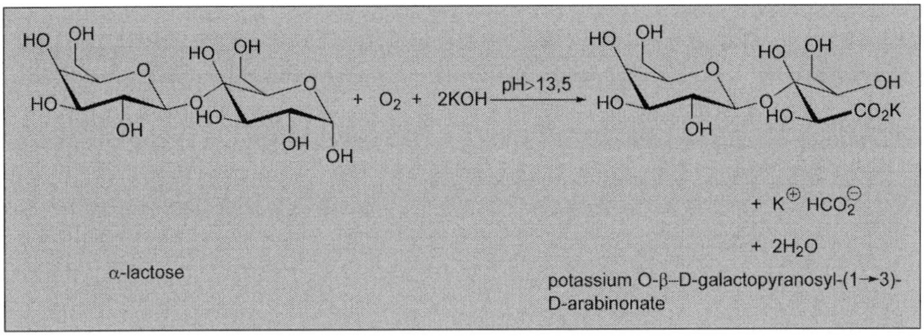

α-lactose

potassium O-β-D-galactopyranosyl-(1→3)-
D-arabinonate

The byproducts are D-galactose, D-lyxonate, 2-deoxy-D-tetronate and glycolate. Adding small amounts of hydrogenperoxide and anthrachinon-2-sulfonate (AMS) increases the selectivity to over 90% and made a higher reaction temperature possible. The role of peroxide and AMS is illustrated below.

[1]

[2]

[3]

[4]

Adding peroxide and AMS separately has no effect.

In aqueous alkaline hydrogenperoxide solutions aldoses are broken down step by step into finally almost quantitatively formic acid. Isbell et al. (8) elucidated this conversion.

Apparently this reaction plays no role in the peroxide/AMS catalyzed oxidation. Van der Berg et al. (9) found that the degradation of aldose with peroxide can be stopped after the first reaction step by adding borate-ions.

Anthrachinons do have an oxidative history in the cellulose chemistry. Holton (10) showed that adding small amounts of anthrachinon to an alkaline wood pulp considerably increased the delignification thereby increasing the yield of pulp.

Cochaux (11) concludes that adding small amounts (1%-2%) anthrachinon (AQ) to an alkaline cellulose dispersion reduces the peeling reaction. This stabilization is connected with the oxidation of the reducing end groups.

$$AQ + 2e^{\ominus} + 2H^{\oplus} \longrightarrow H_2AQ$$

$$\text{cellulose endgroup} - \underset{\underset{O}{\parallel}}{C} - H + H_2O \longrightarrow \text{Cellulose} - C\underset{OH}{\overset{O}{\diagup}} + 2e^{\ominus} + 2H^{\oplus}$$

In the presence of lignin H_2AQ is oxidized under simultaneous depolymerisation of lignin. When using higher concentrations of AQ (>5%) C-6 oxidation besides end group oxidation occurs. Beta elimination causes chain degradation. As stabilization of the end groups also occurs the net result is not a change in the average molecular weight, but an increase in the 'polymolecularity'. Arbin (12) investigated a similar reaction with amylose.

A patent of Roquette (13) describes some interesting details of the peroxide/AMS reaction. During the reaction no excess oxygen is present as a result of effective stirring and adding the oxygen in such a time that the reaction mixture will not become colorless (the reduced form of AMS is red). Sugars are added continuously to the reaction mixture with specific amounts of AMS and alkali calculated with respect to the sugars. This results in less degradation. The oxidation of glucose, glucose syrup, sorbose and carboxymethyl starch is described. The oxidation of sorbose results in xylonic acid with a yield of 75%. The presence of a methanol or redox promoters like hydrochinon or resorcinol enhances the reaction velocity and the yield. The products are proposed as Ca sequestrants.

Another patent of Roquette (14) describes the Spengler- and Pfannenstiel reaction with 5-ketogluconate again using peroxide/AMS as an additional catalyst. The main resulting product is xylaric acid (salt) besides various other low molecular weight organic acids like formic acid, glycolic acid etc.

Some years ago the classical Spengler- and Pfannenstiel reaction with 2-ketogluconate was pointed out:

2-ketogluconate → O₂ → oxalate + C4 aldonate

Both 2- and 5-ketogluconate can be obtained by the oxidation of glucose. Also 2,5 diketogluconate and 3-ketogluconate can be used as substrate for the reaction.

Sakharov et al. (15) describe the catalytic oxidation of polyols by molecular oxygen under alkaline conditions. 'Real' Spengler- and Pfannenstiel reactions cannot occur with polyols like sorbitol - no ene-diol possible -. The use of Cu^{2+} ions as catalyst in the presence of NaOH, KOH and $Ca(OH)_2$ is crucial. The distribution of the oxygenated products of a 20% polyol solution in water is presented in Table 3.

Table 3. Distribution of oxygenated products of a 20% polyol solution $CuCl_2.2H_2O$ 5.10^{-3} mol per l, Temp 50 °C.

Substrate		Mol/l	$\Delta(O2)$ mol/l	Products		
				HCOOH	$HOCH_2COOH$	C_3-C_5 acids
Mannitol	NaOH	0.5	0.34	0.27	0.21	Traces
Mannitol	$Ca(OH)_2$	0.3	0.30	0.23	0.14	0.25
Sorbitol	NaOH	0.5	0.30	0.25	0.22	Traces
Sorbitol	$Ca(OH)_2$	0.3	0.33	0.19	0.17	0.25

The difference in results when $Ca(OH)_2$ is used is remarkable. The initial velocity of the oxygen uptake is much higher than when NaOH, KOH and $Ba(OH)_2$ are used. As $Ca(OH)_2$ is a weak base probably here another reaction mechanism plays a role. Sakharov mentions:" *thus the significant difference observed in product distribution of polyol oxidation with NaOH and $Ca(OH)_2$ as the bases may be due to the formation of more rigid structures of the Cu^{2+}.. A-.. Ca^{2+} complex. This rigidity prevents deep polyol destruction and results in the formation of high molecular acids in considerable amounts".* With NaOH as well as with $Ca(OH)_2$ as catalyst the formation of oxyperoxides is assumed.

Gluconic acid is converted faster than sorbitol into low molecular products. Carbohydrates with a cyclic structure (saccharose, methylglucoside) are only degraded at higher temperatures. The catalytic system $[Cu^{2+}$..base..O] is possibly suited for the degradation of polysaccharides. The kinetics of the starch oxidation are similar to those of the sucrose oxidation.

Especially from the patent literature it is known that $Ca(OH)_2$ can be used instead of NaOH in various alkali catalyzed etherifications of starch. Especially when using $Ca(OH)_2$ in the presence of (air) oxygen, degradation will occur resulting in lower viscosities of aqueous solutions of the products. As etherification of starches is also carried out under (semi-) dry conditions at higher temperatures in the presence of oxygen, the degradation here also is an issue. Strlic et al. (16) describe the degradation of pullulanes with a narrow molecular weight distribution, by oxygen at 80 °C, 65% RH in the presence of $Ca(HCO_3)_2$. An ad random degradation was observed with no cross-link reactions. A correlation was found between the concentration of the reducing end groups and the velocity of degradation and the content of peroxide intermediates after a pre-oxidation treatment. From this it was concluded that the aldehyde groups, via the peroxide intermediates, are decisive for the velocity of the degradation of the polysaccharides.

References

1. Stefan Arts, Thesis. Delft University of Technology. 1996; 20.
2. O. Spengler, A. Pfannenstiel, Z. Ver. Dtsch. Zuckerind. 1935; 85: 546.
3. EP 0755944 (1979) Cerestar.
4. Starch, Chemistry and Technology. Academic Press. 1965; 521.
5. EP 0548399 (1993) Novamont.
6. WO 9218542 (1992) Novamont.
7. H.E.J. Hendriks, Thesis. Eindhoven Technical University. 1991.
8. H.S. Isbell, Carbohydr. Res. 1976; 49: C1.
9. R. v.d. Berg et al., Carbohydr. Res. 1995; 267: 65.
10. H.H. Holton, Pulp Paper Canada. 1977; 10: 218.
11. A. Cocheaux et al., Revue. 1996; 50(5): 191.
12. F.L.A. Arbin, Tappi. 1980; 63(4): 152.
13. FP 2722200 (2004) Roquette.
14. EP 0753503 (1997) Roquette.
15. A.M. Sacharov et al., Kinetics and Catalysis. 1996; 37(3): 368-376.
16. M. Strlic et al., Carbohydr. Polym. 2003; 4: 221-228.

Spacers

In general the industrial preparation of low substituted industrial starch derivatives - ethers and esters - is carried out in water either in aqueous solution or slurry or semi-dry.

Although several hundreds of reagents are described in the literature only about twenty different reagents are used Table 4.

Table 4. Chemical derivatization, reagents used by industry.

Type	Reagent	Effect
• Acetates	Acetic anhydride	Stability of viscosity
	Vinyl acetate	Clear solution
• Hydroxyethyl ether	Ethylene oxide	Stabilization of viscosity
		Clear solution
• Hydroxypropyl ether	Propylene oxide	Stability of viscosity
		Clear solution
• Cyanoethyl ether	Acrylonitrile	Stabilization of viscosity
		Clear solution
• Carboxymethyl ether	Monochloroacetic acid	Increased viscosity
		Clear solution
		Water absorption
• Cationic ether	N-(3-chloro-2-hydroxypropyl)	Positive charge
	Trimethylammoniumchloride	Stabilization of viscosity
• Cross-linked derivatives	Epichlorohydrin	Increased viscosity
	NaTMF	Shear stable
	$POCl_3$	pH stable
	Adipic acid	Temperature stable
		Structure
• Oxidized derivatives	NaOCl	Decreased viscosity
	H_2O_2	
• Hydrolyzed derivatives	H^+	Decreased viscosity
		Gelling
• Dextrins	H^+	Low and stable viscosity
• Phosphates	Phosphoric acid/urea	Stabilization of viscosity
	Interaction with Al-ions	

New products are mostly developed by using combinations of etherification and cross-linking, etherification and oxidation etc. Since the DS values can be varied, there are a large number of products possible.

With the alkali catalyzed reactions there will be mainly C_2-OH substitution (kinetic control) caused by the presence of the relatively acid C_2-OH group (pK$_a$ 12.4). Under acid conditions and with 'bulky substituents' there will be more C_6-OH substitution.

If we are able to carry out the region specific substitution of glucose units, amylose- and amylopectin molecules and specific areas on and in the starch granule, we might obtain derivatives with improved properties. However, we are missing the preparative methods for this purpose.

A new approach to new products is the introduction of a spacer arm between the starch polymer and the functional group. The effect of the functional group is enhanced.
Some examples:
- Galactosyl-β-cyclodextrin derivatives are recognized by galactose specific cell wall lectins. This recognition is dependant on the length of the spacer chain between the cyclodextrins and the sugar head (1).
- The introduction of a short spacer leads to a fast, efficient and region selective glycosylation of β-CD (2).

- The activity of microsphere immobilized glucoamylases is improved by binding the enzyme with a spacer to the microsphere (3).
- Cinnamic acid derivatives of polysaccharides can be cross-linked by radiation with UV-light. The effectivity of the cross-linking can be improved by binding the cinnamic acid group with a spacer to the polysaccharide (4).
- Associative thickeners based on cellulose can be obtained by adding to the aqueous soluble hydroxyethyl cellulose molecules strong hydrophobic long chain alkyl groups. Improvement is achieved by first adding a spacer to the cellulose (5).
- The absorption properties of hydrogels based on cross-linked carboxymethyl cellulose and hydroxyethyl cellulose are improved by adding a spacer. This is done by reacting polyethyleneglycol with two molecules divinylsulfon. In the resulting cross-linker the distance between two cross-link sites is varied by using different molecular weight polyethyleneglycol (6).
- Hydrogels based on carboxymethyl starch were made by heating a derivative with a DS of 0.45 with various organic acids like succinic acid, hydroxyl succinic acid, dihydroxy succinic acid citric acid, maleic acid and adipic acid. The different cross-linkers showed different influence on the strength of the hydrogels. This difference is explained by the difference in spacer length and the number of carboxylic groups (7).

The flexibility of a long polymer chain is limited. The flexibility of any functional group can be improved by incorporating a spacer arm.

The application of spacers is still uncharted terrain. There are interesting possibilities for cationic, anionic, unsaturated and hydrophobic groups bound via a spacer to a polymer as well as with enzymatic conversions of carbohydrates.

References

1. R. Kassab et al., Tetrahedron Letters. 1997; 38(43): 7555-7778.
2. V. Bonneta et al., Tetrahedron Letters. 2003; 44(50): 987-989.
3. M.Y. Arica et al., J. Appl. Pol. Sci. 2001; 81: 2702-2710.
4. USP 6,025,444 (2000).
5. USP 6,248,880 (2001).
6. A. Sinino et al., J. Appl. Pol. Sci. 2003; 90: 168-174.
7. C. Seidel et al., Starch/Stärke. 2001; 53: 305-310.

The phosphate group in sugar- and starch phosphates as nucleophile

Starch phosphates are industrially produced in a dry reaction of starch with phosphates with or without urea present. Using special reagents it is also possible to carry out this phosphorylation in aqueous systems (1). Khalil et al. (2) studied more recently the etherification of starch with the reaction product of monosodium phosphate and epichlorohydrin.

$$NaH_2PO_4 \ + \ H_2C-\underset{\underset{O}{\diagup\diagdown}}{C}-CH_2Cl \ \longrightarrow \ ClH_2C-\underset{\underset{HO}{|}}{C}-\underset{H_2}{C}-O-\underset{\underset{O}{||}}{\overset{\overset{OH}{|}}{P}}-O-Na$$

The resulting product is used in the etherification of starch and thus introduces phosphate groups.

In the starch phosphates the phosphate group is introduced to increase the water binding capacity (thickening agents) or to reduce the solubility in the presence of Al^{3+} ions (paper industry). Another interesting use is as a new nucleophilic centre.

Avison (3) describes the acetylation of α-D-glucose-1-phosphate (G-1-P) with acetic acid anhydride in water at 0°C and pH 7-8. Under the described reaction conditions an acetyl phosphate was formed without reaction of the hydroxyl groups of the sugar.

Also reactive epoxides compounds can react specifically with the phosphate group in G-1-P. The reaction of G-1-P with N-(glycidyl)-trimethylammonium-chloride is patented (4)

$$Glucose-1-O-\underset{\underset{OH}{|}}{\overset{\overset{O}{||}}{P}}-O-\underset{H_2}{C}-\underset{\underset{OH}{|}}{\overset{H}{C}}-\underset{H_2}{C}-\overset{\oplus}{N}(CH_3)_3\overset{\ominus}{Cl}$$

The resulting compound is used in humectants. When one of the methyl groups in the above compound is exchanged for a stearyl group the resulting products can be used in shampoos, household detergents and cosmetics (5).

If the reaction of the phosphate group in G-1-P is selective with suitable reagents it is to be expected that the phosphate groups in starch phosphates will also react selectively. In a patent (5) this reaction is described: "50 g starch phosphate was treated with 19 g glycidyltrimethylammoniumchloride (GMAC) in water for 15 hours to give starch - poly(3-NNN-trimethylamino)-2-hydroxypropyl phosphate". The products are used in skin cleaning compositions.

AKZO (6) has patented the preparation of "hydrophobically modified polyols" starting with polysaccharide phosphates with a molecular weight of more than 1000, in particular polysaccharides such as cellulose, starch and inulin. The phosphates are prepared in the usual way by heating of the polysaccharides with tripolyphosphate. The hydrophobation can be carried out with all kind of possible hydrophobic epoxides - for instance phenyl- and allylglycidylether - by heating the phosphate in an alcohol such as ethylalcohol or isopropylalcohol (no catalyst!). In the described examples exclusively cellulose and inulin are used. It is also mentioned that after the reaction of one hydroxyl group of the phosphate the reactivity of the second hydroxyl group is increased causing an easy binding of two hydrophobic groups to one phosphate. The products are used in various applications such as drilling muds, coating, ink, detergents and cosmetics.

There are advantages for derivatization via the phosphate group. Since the reaction is carried out under neutral conditions there will be little discoloration, which is especially important when starting with reducing oligosaccharides. Etherification under alkaline conditions leads mostly to C_2 substitution. Starting with a phosphate there will be more C_6 substitution and additionally the hydrophobic group is bound via a spacer. As the phosphate group apparently can bind two hydrophobic groups per molecule the hydrophobation is locally higher than with the usual hydrophobation.

Potato starch has the advantage of containing small amounts of phosphate bound as an ester. Also with these groups as a nucleophile the above mentioned reaction will occur. As the phosphate groups are specifically located on the amylopectin molecules there will be a specific amylopectin derivatization.

The reaction of the phosphate group in polysaccharides with other reagents than epoxides is also an interesting topic.

References

1. E.B. Solarec, Modified Starches, Properties and Uses. CRC Press. 1987; 97.
2. M.I. Khalil et al., Starch/Stärke. 2002; 54: 132-139.
3. A.W.D. Avison, J. Chem. Soc. 1955; 732.
4. DE 3,523,835 (1998) KAO co.
5. EP 514,588 (1992) KAO co.
6. JP 05132695 (1993).
7. WO 01/38398 A1 (2001) Akzo.

Acetoacetylation of starch and cellulose

The introduction of keto-groups in starch can be achieved in two ways:
- the oxidation of secondary hydroxyl groups of the starch molecule;
- the introduction of substituents with keto-groups.

As has been pointed out spacers can increase the effectivity of substituents, therefore an external keto-group like in acetoacetates might be preferred over a keto-function in the starch molecule. The acetoacetic acid group makes various reactions possible such as aldol like conversions, Michael additions, reactions with amines etc. Reaction conditions such as UV-curing also then become possible.

Several reactive lactones can react under mild conditions with starch. Diketene reacts with starch in aqueous slurry at pH 7-9 and at temperatures over than 35 °C and forms acetyl acetic acid esters.

The DS values of the reaction product are determined by saponification (1, 2). Especially at low DS values the yields are high (up to 75%). The advantage of the slurry reaction is the fact that the purification of the products can be carried out in a simple way by filtration and washing with water. If more than 25% acetoacetyl is introduced, the dispersability in water of the derivatives diminishes and at over 55% acetoacetyl the products can be made thermoplastic with suitable plasticizers. The reaction of starch in glacial acetic acid and diketene with phosphoric acid as a catalyst is also described as is the preparation of mixed esters (acetoacetate - acetate). The application of the described products is mainly determined by the use of the reactive substituent. The application of starch acetoacetate/formaldehyde is patented for the production of water proof glues (3).

A Japanese patent (4) describes the dry acetoacetylation of starch with diketene. *"500 g corn starch is mixed with 0.3% sodium acetate. On the mixture subsequently 30 parts of diketene is sprayed in 30 min, heated to 80 °C and kept at this temperature for 3 hours".*
The resulting product contains 6.9 mol% acetoacetyl.

Arranz et al. (5) describe the modification of amylose. Amylose dissolved at 80 °C in DMSO was treated with and without a catalyst with diketene. When pyridine was used as catalyst short reaction times resulted and a high DS. The activation energy was determined at a value of Ea =27.2 kJ/mol. The products were characterized with IR, ^{1}H-NMR and ^{13}C-NMR. The DS was determined by titration. With sodium methoxide in benzene and the polymer dissolved in DMF DS values of up to 2.5 were possible. Cu-complexes of the esters were prepared by treating a solution of the ester in THF or a mixture of THF and water with a surplus of Cu acetate in water. A green precipitate appeared. The relative reactivity of the hydroxyl groups of the amylose was determined with ^{13}C-NMR (6). From Table 5 it can be seen that there is relatively more C_6 substitution. This could be caused by steric effects. Also the reactivity of the C_3-OH groups is remarkable.

Table 5. DS of acetoacetylated amylose as a function of the OH group.

DS tot	DS C_2	DS C_3	DS C_6
0.52	0.11	0.15	0.25
1.08	0.23	0.28	0.56
1.76	0.39	0.53	0.84
2.42	0.66	0.91	0.89

In view of the negative effects of diketene - lachrymatory, very reactive, instable and toxic - alternative acetoacetyl reagents were sought. Gotlieb researched the use of methylacetoacetate. He heated potato starch in a steel vessel with this reagent for some hours at 140 °C under vacuum. As the evaporated reagent was not recovered the yield was low (25%).

Cellulose

Edgar et al. (7) describe the preparation of cellulose acetoacetates. He states that based on the literature about the reaction of cellulose with diketene to acetoacetates, it can be concluded that the heterogeneous acetoacetylation causes problems and that the preparation of products with a high DS is difficult. He opted for a reaction in a homogeneous environment. The esterification of the cellulose was carried out in N,N-dimethylacetoamide (DMAC) /LiCl and 1-methyl-2-pyrrolidinone (NMP)/LiCl without catalyst. At 110 °C the reaction was complete in 30 min. The end products were isolated by flocculation in methanol and washing by methanol extraction in a Soxhlet. Crucial in the preparation is the removal of small amounts of water prior to the reaction by the azeotropic distillation of small amounts of solvent. This made yields of 85% - 90% possible with very high DS. Besides diketene, t-butylacetoacetate is also a useful reagent. Firstly acetyl ketene is formed which reacts with cellulose to acetoacetate.

In the used procedure the acetylation with anhydride - without catalyst - was possible. Mixed esters - like acetoacetates/acetates - can be prepared in a simple way. The mixed esters are better soluble than the acetoacetates. The solubility of such mixed esters is less than those of acetobutyrates. By varying the ratio of the DS in the mixed esters it is possible to influence the T_g. Low substituted products (DS<0.5) turned out to be water soluble. The DS values were determined with NMR. Such water soluble products might find application in water borne coatings. The potential for cross-linking of cellulose acetoacetates/acetates with hexakis(methoxymethyl)melamine and α,ω-bis(amino) capped poly(oxypropylene) was investigated. In the case of melamine para-toluenesulphonic acid was used as catalyst. With the obtained solution metal plates were coated and cured. The resulting cross-linked films appeared promising as they showed excellent color, high hardness and good resistance against rubbing with 2-butanone and DMF.

In a German patent (8) the conversion of cellulose with various lactones in DMAC/LiCl catalyzed with bases is described. An example mentions the conversion with lactide.

Diamantoglou and Kuhne (9) described various other esterifications and etherifications of cellulose with DMAC and NMP/LiCl. They mention advantages of the homogeneous derivatization as:
- precise DS;
- homogeneous substituent distribution;
- high yield;
- low amount of by products;
- more possibilities for new derivatives.

Other reaction media for the homogeneous derivatization of cellulose they mention are:
- N_2O_4/DMSP/, DMF or pyridine;
- formaldehyde/DMSO, DMF, NMP;
- quaternary ammonium salts such as N-ethyl pyridinium chloride;
- amine oxides such as N-methylmorpholin-N-oxide.

Besides the already mentioned t-butylacetoacetate as a precursor for acetylketene 2,2,6-trimethyl-4-H-1,3-dioxin-4-on can also be used as an alternative for diketene as acetoacetylation agent. This reagent decomposes at temperatures over 100 °C in acetone and acetylketene (10).

Polyvinylalcohol and other alcohols

In the literature (11, 12) the preparation of acetylacetic acid esters of various (poly)alcohols is also described. In the preparation of polyvinylalcohol-acetoacetates, diketene is used; the transesterification of β-ketoesters is also mentioned.

Alkyl-2-acetoxycrotonates and alkyl-2-acetovinylacetates are used in the acetoacetylation of aliphatic alcohols (50°C - 100°C) in the liquid phase (13).

$$R-OH + H_3C-\underset{\underset{O=C-CH_3}{|}}{\underset{|}{\overset{O}{\overset{\|}{C}}}}=C-\overset{O}{\overset{\|}{C}}-O-CH_3 \longrightarrow H_3C-\overset{O}{\overset{\|}{C}}-\overset{H_2}{C}-\overset{O}{\overset{\|}{C}}-O-R + CH_3OH + CH_3COOH$$

The preparation of glycerol monoacetoacetate is also of importance in this context (14).

$$\begin{array}{c} H_2C-OH \\ | \\ HC-OH \\ | \\ H_2C-OH \end{array} + H_3C-\overset{O}{\overset{\|}{C}}-\overset{H_2}{C}-\overset{O}{\overset{\|}{C}}-O-C_2H_5 \xrightarrow[65-75^\circ C/t=3u]{C_2H_5ONa} \begin{array}{c} H_2C-OH \\ | \\ HC-OH \\ | \\ H_2C-O-\overset{O}{\overset{\|}{C}}-\overset{H_2}{C}-\overset{O}{\overset{\|}{C}}-CH_3 \end{array}$$

The classical synthesis of acetyl acetic acid esters is given as

$$H_3C-\overset{O}{\overset{\|}{C}}-O-C_2H_5 + H_3C-\overset{O}{\overset{\|}{C}}-O-C_2H_5 \xrightarrow[2)\ H^\oplus]{1)\ NaOC_2H_5} H_3C-\overset{O}{\overset{\|}{C}}-\overset{H_2}{C}-\overset{O}{\overset{\|}{C}}-O-C_2H_5$$

In an analogous way starch- and cellulose acetates and ethyl acetates should be converted with sodium ethanolate (not an easy synthesis). The esterification of alcohols catalyzed with lipases is now a well known technology (15). Yadav et al. describe the transesterification of methylacetoacetates with n-butanol catalyzed with lipase (16). This might give possibilities for the preparation of acetoacetates of sugars and polysaccharides.

Unsaturated acetoacetyl monomers have been used in the preparation of co-polymers which can be cross-linked without setting free formaldehyde. The products are used in paper making and give a good dry and wet strength to the paper (17). Also in the preparation of starch esters and starch graft polymers this type of monomer can be used.

Other examples of applications of polyvinylalcohol acetoacetates are:

- water proof films based on glyoxal (18);
- granular cross-linked gel: PVA-acetoacetate+sodium-alginate+$CaCl_2$ +UV curing (19);
- water proof fibers/sheets based on polyvinyl alcoholacetoacetates +UV curing (20);
- thermosetting coatings (21);
- cross-linked hydrogels with aldehydes and hydrazide for enzyme immobilization (22);
- granular cross-linked gels based on PVA acetoacetates and dialdehyde starch (23);
- acetoacetylated PVA resins for gels, photosensitive materials, chelate exchange resins, molding materials and paper reinforcing (24).

References

1. USP 2,654,736 (1953) National Starch.
2. NOA 6,610,306 (1967) National Starch.
3. USP 3,361,585 National Starch.
4. JP 8,11,902 (1981).
5. F. Arranz et al., Makromol. Chem. 1986; 187.
6. F. Arranz et al., Polymer. 1987; 28: 1829.
7. K.G. Edgar et al., Macromolecules. 1995; 28: 4122-4128.
8. DE 3,322,118 (1983).
9. M.Diamantoglou, H.Kuhne, Das Papier. 1988; 42: 690.
10. W.P. Pawloski et al., Carbohydr. Res. 1986; 156: 232-235.
11. JP 80157641 (1980).
12. C.G. Berringer et al., J. Appl. Pol. Sci. 1963; 7: 1797.
13. EP 43088 (1982) Hoechst.
14. JP 02172949 (1990).
15. DE 4,329,293 (1995) BASF.
16. Ganapati D. Yadav et al., J.Mol. Catalysis B. 2005; 32(3): 107-113.
17. EP 458.561 (1991) Vinamul.
18. JP 80157641 (1980).
19. JP 01118529 (1989).
20. JP 6352880 (1988).
21. JP 63243399 (1988).
22. C. Robert et al., J. Coating Technol. 1989; 61(770): 83-91.
23. K. Massao et al., Hakko Kogaku Kaishi. 1991; 69(5).
24. JP 63142031 (1988).
25. JP 044772 (1979).

Derivatization with mixed anhydrides

Commercially available starch esters are predominantly the starch acetates. There is an extensive literature (1) on the preparation and application of starch esters. In this preparation mostly the reactive derivatives of organic acids such as anhydrides, acid chlorides, esters, lactones etc. are used. At a relatively low DS the modification is sufficient to retard or eliminate retrogradation.

Low substituted starch acetates are industrially prepared by reaction of starch with acetic acid anhydride in aqueous alkaline slurry. This acetylation is a two step reaction. First starch reacts with hydroxide after which the acetylation occurs.

The starch nucleophile adds onto one of the two carbonyl groups with an acetate ion as the leaving group (2). In Figure 7 the acetylation of potato starch in aqueous suspension with acetic acid annydride as a function of the pH is shown.

Higher substitution can be achieved in the absence of water using organic solvents. A German patent (3) describes this preparation in water-free N-methylpyrrolidone in the presence of catalysts like 4-dimethylaminopyridine.

The preparation of starch esters of higher carboxylic acids via the anhydride becomes more difficult with increasing number of carbon atoms. This already starts with the anhydride of butyric acid.

When esterifying with *mixed* carboxylic acid anhydrides there are two electrophilic carbonyl groups involved with different reactivity as well as two different leaving groups. The composition of the reaction products differ from case to case. In

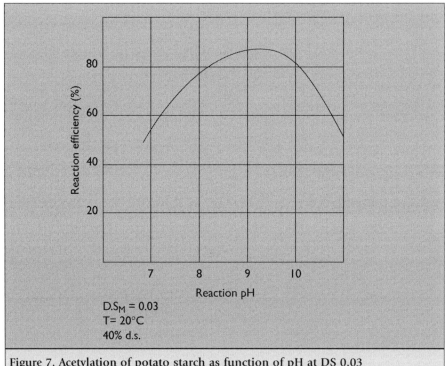

Figure 7. Acetylation of potato starch as function of pH at DS 0.03

an old German patent (4) the preparation of mixed esters is described. The products are prepared with the mixed anhydrides of acetic acid / butyric acid and acetic acid / propionic acid. The reaction is carried out in methylene chloride with an acid catalyst. The mixed anhydrides are prepared by reacting butyric acid etc. with ketene. In a patent of National Starch (5) the reaction with following mixed anhydride is described.

This mixed anhydride is prepared from the acid by conversion with acetyl chloride. After reaction with aqueous alkaline starch slurry the derivative is formed. After hydrolysis of the acetal groups into aldehyde groups the product is cross-linked. This product is used to improve the water resistance of paper. Whether or not any acetylation of starch occurs during the reaction with the anhydride is not clear.

If the mixed anhydride of acetic acid / trifluoroacetic acid is used acetylation almost exclusively occurs. In this way amylose is acetylated in a mixture of one volume glacial acetic acid in two volumes of trifluoroacetic acid. It is assumed that trifluoroacetic acid acetate (see below) is formed (6).

$$F_3C-\overset{\overset{O}{\|}}{C}-O-\overset{\overset{O}{\|}}{C}-CH_3$$

The reaction is thought to be of the type S_N2 addition - elimination or S_N1. Roberts inclines to the latter as in cellulose acetate (DS 2.5), the content of trifluoroacetate groups is minimal and the polarity of the reaction medium is sufficient to produce a charge separation (trifluoroacetic acid is supposed to solvatate the trifluoroacetic acid ion). Hence, there exists equilibrium between trifluoroacetic acid and the acetylium ion and trifluoroacetate:

$$H_3C-\overset{\overset{O}{\|}}{C}-O-\overset{\overset{O}{\|}}{C}-CF_3 \rightleftharpoons H_3C-\overset{\oplus}{C}=O + \overset{O}{\underset{\ominus O}{}}C-CF_3$$

Next the acetylium ion can react with amylose by nucleophilic substitution.

$$\text{amylose}-OH + \overset{CH_3}{\underset{\overset{\|}{O}}{\oplus C}} \longrightarrow \text{amylose}-O-\overset{\oplus}{\underset{OH}{C}}-CH_3 \longrightarrow \text{amylose}-O-\overset{\overset{\|}{O}}{C}-CH_3 + H^{\oplus}$$

Formic acid anhydride is contrary to acetic acid anhydride an instable compound. The mixed anhydride of formic acid and acetic acid can be prepared by reaction of formic acid and ketene (7). Upon storage a disproportionation occurs resulting in formic acid anhydride and acetic acid anhydride. Formic acid anhydride and acetic acid anhydride could be present in mixtures of acetic acid anhydride and formic acid. According to Wu et al. (8) the mixed anhydride of formic acid/acetic acid anhydride is more similar to formic acid anhydride than to acetic acid anhydride. Alcohols could be esterified with mixtures of acetic acid anhydride and formic acid in pyridine resulting predominantly into formiates. Formic acid esters of starch are now prepared by heating starch in the presence of formic acid.

Chloroacetic acid has been used as an "impeller" in the esterification of cellulose and starch with higher fatty acids. This is illustrated by following example:
"In a reaction vessel 30 parts of stearic acid, 6 parts of acetic acid, 50 parts of chloroacetic acid, 50 parts of cellulose and 0.05 parts of Mg perchlorate are reacted for 7 hours."

It is well possible to prepare in this way mixed esters. The formation of chloroacetic acid mixed anhydrides is obvious.

The reaction of starch with the mixed anhydrides of acetic acid and dicarboxylic acids such as adipic acid is interesting. This anhydride plays a role in the industrial preparation of cross-linked starches (9). By heating adipic acid in acetic acid anhydride the following reactions take place:

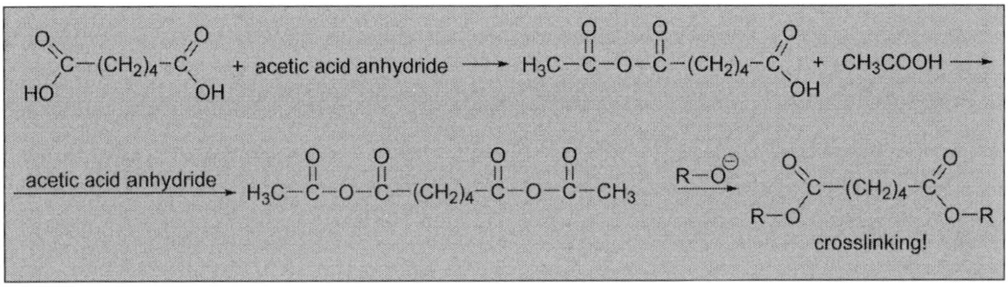

Several side reactions also occur, besides cross-linking there is acetylation, also carbonyl groups are introduced via the mono anhydride. This reaction has never been studied in detail.

National Starch describes in an other patent (10) the acetylation with acylphosphates.

$$R-COO-\overset{\overset{\displaystyle OM_2}{|}}{\underset{\underset{\displaystyle O}{||}}{P}}-OM_1$$

R= alkyl or substituted alkyl, M=metal ion

Succinyl -, chlorendyl-, phtalo-, tetrachlorophtaloyl - and tetrabromophtaloyl-monophosphate are mentioned as reagents, surprisingly the more simple acetyl phosphate not.

The reaction of acylphosphate can be carried out in aqueous alkaline starch slurry; phosphate ions are better leaving groups than acetate ions. Similar to acetyl phosphate, also acetyl chloride can be regarded as a mixed anhydride of acetic acid and hydrochloric acid. In principle acetyl chloride is an acetylating agent for starch, however because of the low yield hardly applicable in aqueous alkaline slurry.

Acyl cyanides (see below) can be regarded as mixed anhydrides of a carboxylic acid and hydrogen cyanide and is described for the acetylation of cyclodextrins (11).

$$\text{acylcyanides} \qquad R-\overset{\underset{\underset{\displaystyle O}{||}}{}}{C}-C\equiv N$$

The presence of other reactive groups in a carboxylic acid can lead to complications in the reaction with acetic acid anhydride. As an example the reaction of acetic acid anhydride with anthranilic acid, can be mentioned. Not the expected acetyl anthranilic acid anhydride is formed but acetylanthranyl.

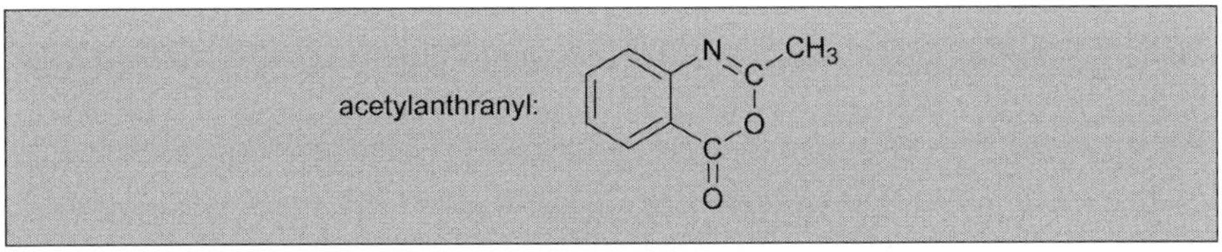

acetylanthranyl:

Acetyl anthranyl reacts with polysaccharides in DMF/pyridine and $NaOCH_3$ as catalyst at 60°C - 150°C and forms acetyl anthranilic acid esters. It is not known if this reaction also occurs in water.

The reaction of an other cyclic mixed anhydride, isatoic anhydride, with starch is described (12). The oxygen analogue of isatoic anhydride is known (13), the reaction with starch is not reported.

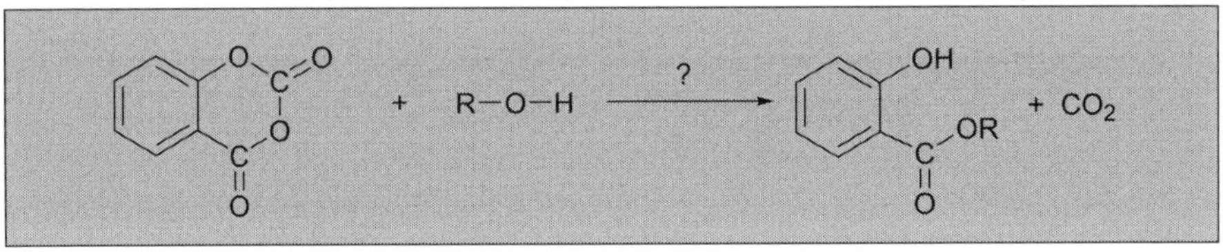

The reaction of α-hydroxy iso-butyric anhydrosulphite (see below) with amines leads to amides (14). The reaction with starch is not known.

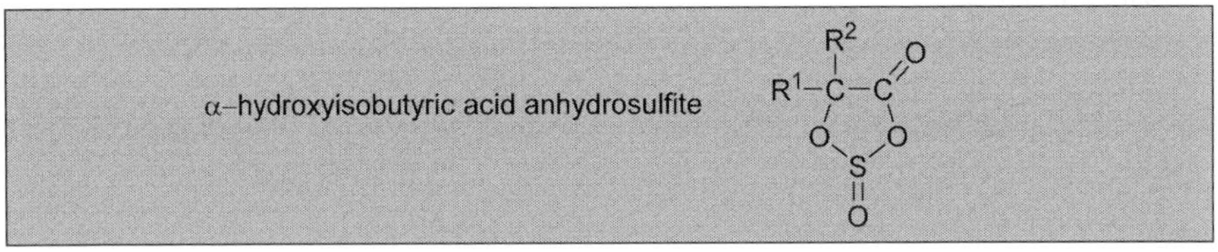

α–hydroxyisobutyric acid anhydrosulfite

Carbonic-carboxylic anhydrides can react with starch slurry under alkaline conditions. Examples have been published (15) where the carbonic anhydrides of acrylic acid, methacrylic acid, propionic acid, lauric acid, stearic acid, benzoic acid and crotonic acid were used. National Starch has patented (16) the cross-linking of starch with the mixed carbonic acid anhydrides of a polycarboxylic acid.

crosslinker: $C_2H_5-O-\overset{O}{\overset{||}{C}}-O-\overset{O}{\overset{||}{C}}-(CH_2)_4-\overset{O}{\overset{||}{C}}-O-\overset{O}{\overset{||}{C}}-O-C_2H_5$

The cross-linking is identical to the reaction with acetic acid/ adipic acid anhydride. The carbonic acid / adipic acid anhydride derivatives are claimed to have a better taste in food applications. The transesterification of alcohols with carbonic acid anhydride / carboxylic acid with lipases as catalyst is reported (17). This presents possibilities for starch derivatization. In a patent of Gevaert (18) mixed anhydrides of carboxylic acids and sulfonic acids are used for the preparation of carbonic acid esters of starch.

Also the reaction of starch acetate with ortho phtalic acid and paratoluenesulfonic acid is described. After 24 hours at 80 °C the reaction product is isolated by precipitation in alcohol. The DS is 0.43 based on the ortho phtalic groups. Also H.E starch ortho phthalate and an acetate (DS 0.5)-propionate (DS0.4) is described.

Benzoyl thiosulphate is an excellent water soluble benzoylating agent for starch (19).

Starch benzoates based on benzoylthiosulfate gelatinize better than products based on benzoylchloride. The reaction of mixed anhydrides of carboxylic acids and aliphatic sulfonic acids with starch might be carried out analogous to the reaction of benzoates and benzoylthiosulfate.

References

1. Modified Starches, Properties and Uses. CRC Press. 1986; 55: 131.

2. Starch, Chemistry and Technology. Academic Press. 1965; 447.

3. DP 4423681 (1996).

4. DP 859445 (1952).

5. EP 0314063 (1988) National Starch.

6. Starch, Chemistry and Technology. Academic Press. 1965; 449.

7. A. van Es, Thesis. University of Leiden. 1964.

8. G. Wu et al., J. Phys. Chem. 1996; 100(28): 11620-11629.

9. USP 2,035,510 (1960).

10. USP 3,928,321 (1974).

11. JP 0712003 (1995).

12. BP 1,190,000 (1970).

13. M.H.Davies, J. Chem. Soc. 1951; 1357.

14. Macromol. Chemie. 1969; 125: 170.

15. USP 3,720,662 (1973).

16. USP 3,699,095 (1972).

17. E.G.Jampel et al., Tetrahedron Letters. 1996; 52(12): 4397-4402.

18. DAS 1174753 (1964).

19. BP 1,461,608 (1977).

Derivatization with alkyl sulfates

Sulfonic acid esters behave differently from carboxylic acid esters during hydrolysis. Sulfonic acid esters cleave at the alkyl oxygen while carboxylic acid esters cleave at the acyl oxygen. Also in reactions with alcohols these differences are observed, esters of sulfonic acid are alkylating agents while carboxylic acid esters can be used for trans-esterification.

Dimethylsulfate is well known as methylating agent for carbohydrates. Haworth described the methylation of sugars in aqueous alkaline solution about a hundred years ago (1). Hall et al. (2) described the selective methylation of the C_1-OH of glucose in aqueous alkaline solution. This selective alkylation has also been used in the preparation of surfactants (3) whereby dialkylsulfates with an alkyl-chain length of 6 to 30 carbon atoms were used. Although it is stated that the reaction of the reducing sugars can be carried out in aqueous alkaline solution only examples of the reaction with didecyl sulfate in organic solvents (hexamethylphosphoric acid triamide en dimethylpropyleneurea) with NaH as a catalyst are described.

Haworth also elaborated the permethylation of potato starch with dimethyl sulfate under alkaline conditions. After the methylation and acid hydrolysis a mixture of glucose ethers is obtained. Quantitative analysis of this mixture produces information on the structure of the starch used. Haworth's method has been improved over the years. Application of a microwave in the permethylation of guar gum with dimethyl sulfate reduced the reaction time and hydrolysis time appreciably (4).

The methylation of native granular potato starch in aqueous suspension for the preparation of low DS derivatives is patented (5). The preparation of potato starch ether with DS 0.06 can be carried out as follows:
"In a vessel of 1 l placed in a water bath at 35 °C 200g potato starch - with 20% moisture - was placed. The pH was raised with a 3% NaOH solution to pH 11. 12.6 g (0.1 mol) dimethylsulfate was added and the pH was kept at 11 during 1 hour with 3% NaOH solution. After neutralizing with HCl to pH 6.0 the reaction product was filtrated over a Büchner funnel, washed till salt free and dried at 40 °C".

These methyl ethers are characterized by a favourable gelatinization behaviour differing hardly from native potato starch. This is caused by the character of the substituent and the mild reaction conditions with no swelling inhibitor, resulting in little annealing.

Burgt et al. (6) investigated the distribution of the methyl substituent in methylated potato starch. Applying extensive α-amylase degradation and precipitation in alcohol two fractions were obtained which corresponded with the branched and linear parts of the amylopectin molecule. After determination of the DS, it appeared that during the methylation of potato starch in aqueous suspension with dimethylsulfate the substitution occurs preferently on the amorphous parts, close to the branching points of the amylopectin molecule. The amylose part showed higher DS than the linear parts of the amylopectin.

During the methylation of potato starch with dimethyl sulfate the monomethyl sulfate ion is formed as a by-product. This does not react with starch. It would be interesting to know under which conditions a reaction might occur. Under acid conditions monomethyl sulfate dissolved in DMSO reacts with starch and a starch sulfate is formed (7).

The methanol formed during the reaction is removed under vacuum. This reaction is similar to the transesterification of carboxylic acid esters.

Diethyl sulfate can also be used for the alkylation of starch in aqueous alkaline suspension. Ethyl ethers are formed. The reaction is slow.

Dimethylsulfate in combination with ethylene imine is used for the cationization of starch. Ethylene imine is quaternized in situ after which reaction with starch takes place (8). The reaction is carried out under the usual conditions for the preparation of a cationic starch derivative.

Methyl chloride can also be used for the methylation of starch, however due to the low reactivity it is not suitable for reactions in aqueous alkaline starch suspensions. Under "dry" conditions at higher temperatures the reaction is faster. On an industrial scale there is no production of starch methyl ethers. Cellulose methyl ethers are widely produced.

Ascorbic acid can be selectively alkylated on the C_3-OH without protecting the C_5 and C_6 OH groups. The reaction is carried out with alkylmesylates in DMSO and $NaHCO_3$ as catalyst at 60 °C during 12 hours (9).

The alkylation of starch with alkylsulfonates is hardly studied. The sulfopropylation of starch in aqueous slurry with propanesulton is known (10). The reactivity of this compound relates probably to the cyclic structure.

Propane sulton

Other reagents for the preparation of sulfoalkylethers are:

$$Cl-\underset{H_2}{C}-\underset{H_2}{C}-\underset{H_2}{C}-\overset{\displaystyle O}{\underset{\displaystyle O}{\underset{\|}{\overset{\|}{S}}}}-O-Na$$

$$Cl-\underset{H_2}{C}-\underset{\underset{OH}{|}}{\overset{H}{C}}-\underset{H_2}{C}-\overset{\displaystyle O}{\underset{\displaystyle O}{\underset{\|}{\overset{\|}{S}}}}-O-Na \quad \text{or the epoxide}$$

$$H_2C=\underset{H}{\overset{}{C}}-\overset{\displaystyle O}{\underset{\displaystyle O}{\underset{\|}{\overset{\|}{S}}}}-O-Na$$

$$Cl-\underset{H_2}{C}-\overset{\displaystyle O}{\underset{\displaystyle O}{\underset{\|}{\overset{\|}{S}}}}-O-Na \quad \text{and} \quad Cl-\underset{H_2}{C}-\underset{H_2}{C}-\overset{\displaystyle O}{\underset{\displaystyle O}{\underset{\|}{\overset{\|}{S}}}}-O-Na$$

BASF patented (11) the preparation of ethylene sulfite and the possibilities for using this reagent for hydroxyethylation.

$$H_2C-CH_2 + SO_2 + H_3C-S-CH_3 \longrightarrow \underset{H_3C}{\overset{H_3C}{\diagdown}} SH-\underset{H_2}{C}-\underset{H_2}{C}-O-SO_2^{\ominus}$$

$$\longrightarrow \underset{\underset{\underset{O}{\|}}{\overset{S}{\diagup\diagdown}}}{H_2C-CH_2} + H_3C-S-CH_3$$

The hydroxyethylation of starch with this reagent has not been reported. The reaction with ethylene carbonate is published (12, 13). Also ethylenesulfate is known as a hydroxy alkylating agent (14). The reaction with (poly)acrylic acid is as follows

O
||
Polyacrylic—C—OH + ethylene sulfate ⟶ Polyacrylic—C—O—C—C—O—S—O⁻ Na⁺ ⟶

(with H₂ H₂ and O=S=O groups)

H⁺
⟶ Polyacrylic—C—O—C—C—OH + NaHSO₄

A similar reaction with starch is suspected but not known.

Massoneau (15) has described the formation of cyclic sulfates from cyclic sulfites. He describes the reaction of a cyclic sulfate with an alcohol. A similar reaction with starch could be interesting.

Ar—OH + H₂C—C—C—Cl (cyclic sulfate) →[NaOH/H₂O, acetonitrile, 20°C] Ar—O—C—C—C—Cl (with O—S—O sulfate group) →[H⁺, 20°C]

Ar—O—C—C—C—Cl (with OH) →[NaOH/H₂O, 20°C] Ar—O—C—C—CH₂ (epoxide)

BASF has patented (16) the preparation of hydroxyl ethyl- and sulfato ethyl glycosides. Monosaccharides are converted in DMF with a basic catalyst.

H₂C—C—C—R (cyclic sulfate)   R - long chain alkyl

reaction product   Gly—C—C—R (with X)   X—OH or sulfate

From the sulphate the hydroxide is obtained by acid hydrolysis. This reagent could have potential for starch derivatization.

In his thesis Unrau (17) describes the synthesis of starch sulfates from starch and a DMSO-SO3 complex in DMSO. With alkali as catalyst at 15 °C a derivative with a DS 1.0 was converted into a product containing 25% 3,6-anhydro-D-glucose units. This is analogous to γ-carrageenan which contains in the precursor units C_6 sulfate groups which can form C_6-C_3 ether bridges with sulfate as the leaving group.

Starch sulfates can also be produced in a slurry reaction with amine-SO_3 complexes, N-imidazole derivatives and fluorosulfates. The low substituted starch sulfates give a clear, viscous and stable paste. Industrially this reaction is not carried out. If a cheap reagent would be available this might change. High substituted derivative are investigated as heparin substitute, pepsin inhibitor and for the treatment of stomach ulcers.

With the exception of p-toluene sulfonic acid esters the sulfonic acid esters of starch have hardly been studied. With pyridine as a solvent, high substituted tosylates can be obtained by reacting starch with p-toluenesulfochloride (18). With suitable reaction conditions C_6 substituted monotosylates can be obtained. Also in aqueous slurry starch can be tosylated. The derivatives are poorly water soluble. Even after removal of most of the tosylate groups for example with an amine, the solubility is a problem. Probably cross-links, possibly oxygen-bridges are formed during the tosylation.

References

1. W.N. Haworth, J. Chem. Soc. 1915; 107: 13.
2. D.M. Hall et al., Carbohydr. Res. 1970; 12: 421-428.
3. WO 9321197 (1993).
4. Vandana Singh et al., Tetrahedron Letters. 2003; 44: 7295-7297.
5. USP 2,858,305 (1958).
6. Y.E.M. van der Burgt et al., Carbohydr. Res. 1998; 312: 201-208.
7. R. Takana et al., J. Carbohydr. Chem. 2000; 19(9): 1185-1190.
8. BP 1,399,143.
9. U. Beifuss et al., Synlett. 1999; 1: 147-149.
10. USP 3,046,272 (1962).

11. DOS 1223397 (1966).

12. USP 4,474,951 (1984).

13. M.C. Srivastava et al., Indian J. Chem. 1971; 9: 1081.

14. M. Storm et al., European Polymer J. 1978; 14: 977.

15. V. Massoneau, New Y. Chem. 1992; 16: 207-212.

16. DOS 4,404,728 (1995).

17. D.G. Unrau, Thesis. Purdue University. 1968.

18. Methods in Carbohydrate Chemistry. Academic Press. 1964; 4: 299.

Cyanamide, a reagent with potential

Calcium cyanamide (See below) is prepared by heating calcium carbide in nitrogen to temperatures over 1000 °C. The product is a blackish powder caused by residual carbon. Calcium cyanamide is used as a fertilizer. Cyanamide can be easily extracted from calcium cyanamide with water.

Aqueous alkaline solutions of cyanamide are not stable. At pH 9.5 dicyanodiamide is formed and at still higher pH's urea. Therefore the reagent has to be used quickly or neutralized with nitric acid or hydrochloric acid.

Cyanamide reacts with nucleophiles such as amines and alcohols. The reaction with methanol is catalyzed by acid and results in isourea derivatives which are strongly basic. The reaction with an amine results in guanidine derivatives.

cyanamide: $H_2N-C\equiv N$

H_3C-OH + $H_2N-C\equiv N \cdot HCl$ ⟶ $H_3C-O-\underset{\underset{NH}{\|}}{C}-NH \cdot HCl$

$R-NH_2$ + $H_2N-C\equiv N$ ⟶ $H_2N-\underset{\underset{NH}{\|}}{\overset{\overset{H}{}}{C}}-N-R$

$Ca{\big\langle}^{C}_{C}$ ⫴ + N_2 ⟶ $CaNCN$ + C

The possibilities for starch derivatization with cyanamide have been extensively studied (1). Usually the reaction is carried out with starch in aqueous alkaline slurry; the conversion under acid conditions is also possible (2). Under acid conditions the reagent is converted into its more reactive protonated form, under alkaline conditions the rather acid C_2-OH groups of the starch dissociate into more nucleophilic C_2-O-ions. In both cases iso-urea substituted products are formed.

On an industrial scale the reaction of starch with calcium- or sodiumcyanamide is carried out in aqueous slurry at pH 10 at 20 °C during 24 hours. After filtration and washing the pH of the slurry is lowered to pH 2 and the slurry is again filtrated and dried to 12% moisture. Although this preparation is simple and cheap it never has been successful. This is caused by the instability of the formed iso-urea starch derivatives; after certain time nitrogen is lost and cross-linking occurs. The products are stable at low pH and low moisture.

Urea reacts with glucose and can form glucosylureide. There has hardly been research into the possibilities to convert the iso-urea substituents in cyanamide starch with suitable reagents. Such a conversion could possibly stabilize the substituents. A possibility is the conversion of the iso-urea groups with reducing sugars, oligosaccharides etc.

With cyanamide glucose can react and forms 2-aminooxazolin (3).

The reaction of cyanamide with amines has been mentioned earlier. It seems possible to convert starch derivatives with free amino groups with cyanamide in such a way that the amino groups react and the hydroxyl groups not. In that case, the weak basic amino groups are converted into the strong basic guanidine groups. Such a starch derivative could be an alternative for the quaternary ammonium substituted cationic starch. A similar conversion of chitosan and of the amino groups in proteins might be considered. The synthesis of kreatinin with cyanamide has been patented (4).

The amino acid arginine contains also a guanidine group. The enzymatic preparation of surfactants based on arginine has been patented (5). The products have excellent emulsifying properties, are anti-microbial and low toxic. Esterification of starch with arginine could result into derivatives with guanidine substituents.

A speculation can be made about the reaction of glucamine, glucosamine etc. with cyanamide as precursor for the preparation of biodegradable cationic surfactants. The hydrophilic part contains besides a guanidine group also a sugar group.

Cyanamide activates hydrogen peroxide. Under alkaline conditions first a peroxyimidic acid is formed which gives various decomposition reactions.

$$H_2N-C\equiv N \quad + \; H_2O_2 \longrightarrow H_2N-\underset{\underset{NH}{\|}}{C}-O-O-H$$

After homolysis urea and hydroxyl radicals are formed. The role of these radicals in reactions with lignin model compounds has been studied (6). The system cyanamide/H_2O_2 could perhaps be used in the oxidation, epoxidation and grafting of starch.

References

1. Modified Starches, Properties and Uses. CRC Press. 1987; 118.
2. USP 3,496,155 (1970) American Cyanamid.
3. Trostberg, Brochure SKW-Cyanamid. L 500.
4. EP 0754679.
5. P. Clapes et al., Biotechnol. Bioeng. 1999; 63(3): 333-343.
6. Kadla, Holzforschung. 1998; 52: 513-520.

Urea and starch derivatization

Urea was first synthesized by Wöhler in 1828 by heating ammonium cyanate. It is used as fertilizer, feed additive and in urea-formaldehyde resins.

$$NH_4CNO \longrightarrow H_2N-\underset{\underset{O}{\|}}{C}-NH_2 \quad \text{urea}$$

Today it is produced by heating liquid ammonia and carbondioxyde under pressure. First ammoniumcarbamate is formed which at lowering the pressure decomposes into water and urea. It is also produced from calcium cyanamide. Urea is a white powder with melting point 132,7°C and easy soluble in water and alcohol. It is stable and strongly polar. Because of the presence of donor- and acceptor sites it can form strong hydrogen bridges. Urea is both a weak base - but more basic than amides like acetamide - and a weak acid the K_b is 10^{-14} (1). Urea forms salts with strong acids which hydrolyse in dilute aqueous solutions.

$$H_2N-\underset{\underset{O}{\|}}{C}-NH_2 + H_2O \longrightarrow H_2N-\underset{\underset{O}{\|}}{C}-\overset{\oplus}{N}H_3 + {}^{\ominus}OH$$

Proteins denaturize in concentrated urea solutions; urea also influences the gelatinization and retrogradation of starch (2).

The literature about the application of urea in starch technology is old and vast. Although the use of urea to prevent phase separation in concentrated partly degraded starch derivative solutions has been well known for a long time there has been recently issued an American patent (3) claiming this purpose. To make the use possible of concentrated (15%- 35%) cationic starch derivatives (M_w 5000 - 500.000) in cosmetic applications substantial amounts of urea are added. Hydroxy acids, like lactic acid or their salts, showed the same effect.

In some dry reactions of starch such as the phosphorylation with phosphoric acid and the dry acid degradation with sulphuric acid the presence of urea is essential. The reason is not completely clear but is thought to be connected with the following functions:

- catalyst / solvent;
- buffer;
- ureide formation;
- carbamate formation.

The reaction velocity of the phosphorylation of starch in urea increases with increasing amounts of urea (4, 5). During the reaction of starch with phosphoric acid, the acid is partly neutralized by the ammonia set free by the decomposition of the urea. This results in less acid hydrolysis of starch. If acid hydrolysis is wanted a stronger acid like sulphuric acid is used. Under acid conditions glucose and oligosaccharides react easily with urea into ureides. This stabilizes the end groups - no Amadori reaction (6) - preventing browning.

$$-C_1-OH \quad + \quad H_2N-\underset{\underset{O}{\|}}{C}-NH_2 \quad \longrightarrow \quad \beta-C_1-NH-\underset{\underset{O}{\|}}{C}-NH_2$$

The hydroxyl groups in starch react easily at higher temperatures and in the presence of acid into carbamate. Possibly the urea first decomposes into isocyanic acid/cyanic acid and ammonia after which the isocyanic acid/cyanic acid reacts with starch into the carbamic acid ester. The boiling point of iso-cyanic acid /cyanic acid is 23.5 °C. Under the usual reaction conditions of 125 °C and vacuum the reagent will have only a short time to react. Probably there is some dimerization of the iso-cyanic acid /cyanic acid causing the formation of starch allophanates. The yield of the carbamate formation based on the amount of urea is low. The effect of urea on color is unique. Acetamide, DMF, formamide, ethylcarbamate and DMSO used as replacement of urea had a negative effect.

$$\text{Carbamic acid ester:} \quad R-O-\underset{\underset{O}{\|}}{C}-NH_2$$

Starch carbamates are also formed without a catalyst when starch is heated with urea at temperatures of 90 °C - 120 °C at 20% moisture. The introduced carbamate groups show an elevated activity for aldehydes making application in paper making possible (7).

Khalil et al. (8) state that the reaction of starch with urea in the absence of an acid catalyst at elevated temperatures and long reaction times leads to cross-linking; ammonia is split off from the carbamate group and forms a carbonic acid ester R-O-C(=O)-O-R Acids like phosphoric acid and sulphuric acid prevent this reaction. The effectivity of the acid catalyst with respect to the carbamate formation is sulphuric acid>nitric acid>phosphoric acid.

There is a vast literature about the preparation of cellulose carbamates with urea. In view of the similarity between cellulose and starch the results can be easily translated to starch. Cellulose is after activation in an organic solvent converted with urea into a carbamate. This conversion is increased if the urea is carefully dosed in distinct portions (9). This reaction can also be carried out continuously (10).

Heating of a mixture of sugar cane bagasse with urea in a microwave leads to chelating agents used for the removal of copper and mercury. The product is as effective as commercially available products (11).

Reaction of starch with the mono- or dimethylol derivatives of urea also results into chelating agents. The methylol derivatives of urea are prepared by reacting urea dissolved in water with 1.1 or 2.2 equivalents of formaldehyde at pH 9 at room temperature for 24 hours. The reaction of starch with the methylol derivatives takes place in a dry form at 150 °C for 10 min. and $MgCl_2$ as catalyst. The absorption capacity of the starch derivative for different ions decreases in the order Hg>Cu>Zn>Co>Cd>Pb. The monomethylol derivatives absorb better than the dimethylol derivatives. Products based on thiourea absorb better (12).

The preparation of high substituted starch phosphates (DS 0.1 - 1) from starch and phosphoric acid in urea has been patented (13). Important for the stability of the product is a DS of at least 0.1 of carbamic acid ester groups. The preparation is described as:
"111.1 g of corn starch with a starch content of 90% and moisture of 10% are mixed with 22 ml water. Under continuous stirring 42.2 g phosphoric acid (85%) and 55.6 g urea is added. The resulting mixture is dried at 90°C under vacuum (25 - 35 mm

Hg). Next the mixture is heated to the desired reaction temperature of 130 - 140 °C
under vacuum for 2 hours.
A dough like product is formed which solidifies when cooled.
The product is milled and cleaned by washing with water/methanol (1:3).
After drying under vacuum at 30 - 50 °C 122.2 g starch phosphate is obtained.
The phosphorous content is 6.6%, nitrogen 2.2%. DS phosphate 0.48, DS carbamide
0.35."

Heinze (14) has reported the preparation of a starch phosphate carbamate with strong swelling properties. Phosphorylation of wheat starch with a mixture of phosphoric acid and urea at 135 °C during 1-3 hours under vacuum resulted in products with a DS phosphate 0.2 - 0.3. The products showed good water binding properties and Cu^{2+} ion absorbing power. Such products which contain (besides phosphate groups) carbamic acid esters, are suited for the purification of waste water and the demineralization of water. An interesting aspect was the influence of the ratio amylopectin to amylose on the DS and physical properties. Both the chemically bound urea as the physically bound urea influences the product properties. Besides phosphate groups also polyphosphate substituents were found. To ensure a good color and conversion rate of the products there has to be at least 3 times more urea present than phosphoric acid. A reaction temperature in the range of the melting point of urea is supposed to be optimal. At higher temperatures biuret might be formed leading to cross-linking. The vacuum ensures a low concentration of ammonia resulting in less color formation.

The reaction of 'waste cellulose" (a misleading name as it is obtained from starch processing residues by acid treatment) with phosphoric acid and urea results in products with super absorbent properties (15).

Gospodinova et al. (16) reported the reaction of micro crystalline cellulose with phosphoric acid and urea using a micro wave.

The reaction of cellulose containing products with phosphoric acid or ammonium phosphate in the presence of urea and elementary sulphur prevents the formation of polyphosphates. This facilitates the purification of the products. Carrying out the reaction under vacuum reduces the reaction temperature to 125 - 145 °C and avoids side-reactions. The products are called "biosorbents" and are used for the removal of heavy metal ions from water (17).

The phosphorylation reaction of other non carbohydrate polymers is hardly described. Hirofusa (18) described the reaction of polyvinylalcohol in DMF with dicyandiamide, urea and phosphoric acid. He reports the formation of phosphorylated polyvinyl alcohol derivatives.

The reaction of β-D-glucopyranosyl ureide by heating glucose and urea in water for 8hrs. at 68°C with sulphuric acid as catalyst is patented (19). The resulting products with a purity of >90% are used in feed. Ruminants can digest urea as replacement of proteins, however too high amounts can lead to ammonia poisoning. β-glucosylureide has more favourable properties as the ammonia is set free in a more controlled way. The glucosylureide also improves the digestion of the hemi-cellulose and cellulose fractions in feed.

The preparation of glucosylureide from glucose and glucose syrups is described in an old patent of O.J.Meijer dextrin factories (20). The products were used as textile auxiliary products for cellulose containing fibers such as cotton as a crease-resistant finish.

Glucose syrups can react under suitable circumstances with urea in water and form end- groups stabilized syrups. These products can sometimes replace hydrogenated syrups (21).

The reaction of D-glucose with urea in an aqueous phenolic solution leads to α-D-glucopyrasonylamine 1,2-(cyclic carbamate) (22).

The reaction of starch with the monomethylol derivative of glucosylureide might result in products with interesting rheological properties.

Urea is also used in small amounts up to 2% in the preparation of phenol-formaldehyde based resins to chemically remove free formaldehyde. Adding more than 4% urea serves to replace phenol to decrease the price of the resin; it also increases the reactivity of the resin. Turunen et al. (23) studied the effects of adding wheat starch, lignin (derivatives) and urea in phenol-formaldehyde resins. He observed exothermic effects during curing.

Starch hydrolysates (mainly glucose, maltose and maltotriose) can also be used in foamed urea-formaldehyde resins (24).

Urea is an auxiliary chemical in bioplastics and carbohydrate based hotmelts (25, 26). Especially in the case of hotmelts the amounts are appreciable. Xiaofei Ma et al. (27) compared various compounds with an amide group as plasticizer for thermoplastic starch. The effects on the mechanical properties and the retardation of the retrogradation are connected to the hydrogen bond forming abilities in the following order urea > formamide > acetamide > polyols.

References

1. P.A.S. Smith, Open Chain Nitrogen Compounds. W.A. Benjamin. 1965; 1.

2. S.T. Erlander et al., Macromolekulare Chemie. 1967; 107: 204-221.

3. USP 6,365,140 (2002).

4. Starch, Chemistry and Technology. Academic Press. 1984; 351.

5. Modified Starches, Properties and Uses. CRC Press. 1987; 10.

6. M.H. Benn et al., J. Chem. Soc. 1960; 3837.

7. Starch, Chemistry and Technology. Academic Press. 1967; 305.

8. M.I. Khalil et al., Carbohydr. Pol. 2002; 48: 255-261.

9. DE 10,040,341 (2001) Zimmer AG.

10. DE 10,102,172 (2002) Zimmer AG.

11. U.S.Orlando et al., Green Chemistry. 2002; 4: 555-557.

12. M.I. Khalil et al., Carbohydr. Res. 2000; 324: 189-199.

13. DE 19859123 C1 (2000).

14. U. Heinze et al., Starch/Stärke. 2003; 55: 55-60.

15. JP 09031101 (1997).

16. N. Gospodinova et al., Green Chemistry. 2002; 4: 220-222.

17. WO 99/28372 (1999).

18. S. Hirofusa et al., Makromol. Symp. 1996; 105: 217-222.

19. NOA 7806274 (1978) Batelle.

20. NO 72875 (1953) O.J.Meijer's Dextrine Fabrieken.

21. Personal communication.

22. R.F. Helm et al., J. Carbohydr. Chem. 1989; 8(5): 687-692.

23. M. Turunen et al., J. Appl. Pol. Sci. 2003; 88 : 582-588.

24. D. Braun et al., Angewandte Makromolekulare Chemie. 1991; 193: 183-194.

25. WO 9311937 (1993) Novamont.

26. EP 609,952 (1994) AVEBE.

27. Xiaofei Ma et al., Carbohydr. Pol. 2004; 57: 197-203.

Starch and sugar derivatives with amidoxime and hydroxamic acid groups

The Michael addition of acrylonitrile to starch introduces cyanoethyl groups.

$$R-OH \ + \ H_2C{=}\underset{H}{C}{-}CN \ \xrightarrow{\ominus OH} \ R-O-\underset{H_2}{C}-\underset{H_2}{C}-CN$$

The reaction is carried out under mild conditions in aqueous alkaline slurry. The reverse reaction and hydrolysis of the nitrile group into an amide and/or carboxylic group hardly occurs (1). Acrylonitrile is well available and relatively cheap. Low substituted derivatives are used in many applications. The cyanoethyl group reduces the retrogradation velocity of aqueous starch solutions and increases the clarity of solutions and the stability of the viscosity. Cyanoethyl ethers gelatinize less easily than hydroxyethyl ethers, they do have a "cross-linked" character. This could be caused by the intermolecular reaction between the OH groups and the nitrile group resulting in amidate esters.

$$R-O-\underset{H_2}{C}-\underset{H_2}{C}-\underset{\underset{NH}{\|}}{C}-O-R$$

This probably explains why cyanoethyl ethers of starch gelatinize more easily under alkaline conditions (ester hydrolysis).

The nitrile groups in cyanoethyl starch inhibit bacterial growth and are responsible for the strong adsorption of cyanoethyl starch to cellulose fibres (2).

Cyanoethyl sugars are known (3). The possibility for the reaction with hydroxyl amine is hardly investigated.

Cyanoethyl starch reacts with hydroxylamine and forms derivatives with amidoxime groups.

starch amidoxime

$$R-O-\overset{H_2}{\underset{}{C}}-\overset{H_2}{\underset{}{C}}-\underset{\underset{NH}{\|}}{C}-NHOH$$

The Kyuritsu Organic Industrial research Laboratory in Japan patented these compounds (4, 5, 6). The amidoxime formation takes place at about 90 °C at pH 8. The reaction time is several hours. The derivatives have a cationic character which makes them suitable for paper making (improvement of strength). Mixed starch ethers with quaternary amino alkyl- and cyanoethyl groups reacted with hydroxylamine are also patented. In paper making applications these compounds demonstrate in comparison with the usual cationic starches a better performance (7).

The amidoxime group forms complexes with heavy metal ions. Cyanoethylated cross-linked starch, reacted with hydroxylamine was able to strip traces of radioactive elements and metals from waste water (8). Lutfor et al. (19) described the grafting of sago starch with acrylonitrile with radical initiators. The nitrile group is converted with hydroxylamine under alkaline conditions. The resulting resin has chelating properties. The order in which metal ions are adsorbed:

$$Cu^{2+}>Fe^{3+}>As^{3+}>Zn^{2+}>Ni^{2+}>Cd^{2+}>Co^{2+}>Cr^{3+}>Pb^{2+}$$

Cross-linked polyacrylonitrile starch graft polymers reacted with hydroxylamine show the same properties. These 'new cationic exchange materials' show a high adsorption for Cu (10).

Amidoxime groups have also been introduced into inulin. Cyanoethyl inulin with a DP 10 is reacted during 5 hours at 95 °C and pH 6.5 with hydroxylamine (11). Cu^{2+} ions form a complex with O-(3-hydroxylimino-3-aminopropyl) inulin. The structure of this complex is shown in the following figure.

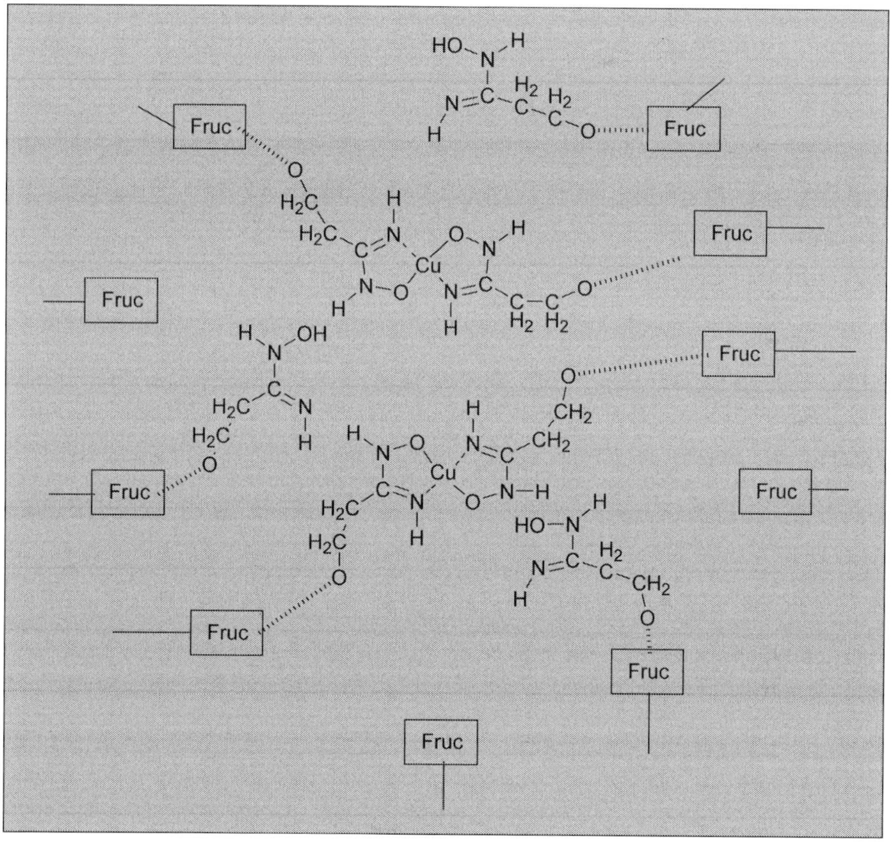

At a ratio Cu ions to amidoxime groups <0.25 2:1 complexes are formed. One Cu^{2+} ion is coordinated by two bidentate amidoxime ligands. The polymer has to adopt such a shape that it becomes possible for two amidoxime groups to approach the Cu ion. Because of steric hindering only half of the amidoxime groups can form stable 2:1 complexes. Increase of the Cu/amidoxime ratio does not lead to more complexing of Cu ions. The complexes are intra-molecular, increase of Cu ions does not lead to viscosity increase.

Graft polymers of polysaccharides, especially cellulose, and acrylonitrile can be used for the preparation of ion exchange resins for the removal of heavy metal ions. A combination of part hydrolysis and part conversion of the nitrile group with hydroxylamine is used (12). Also the reaction of hydroxylamine with bagasse derivatives (13) and with sugar derivatives (3) is described.

A with amidoxime related group of compounds are the hydroxamic acids, the reaction products of carboxylic esters and hydroxylamine.

$$C_nH_{2n+1} - \underset{\underset{O}{\|}}{C} - O - C_2H_5 \quad + \quad NH_2OH \quad \longrightarrow \quad C_nH_{2n+1} - \underset{\underset{O}{\|}}{C} - NHOH \quad + \quad C_2H_5OH$$

It is also possible to start with amides, anhydrides or acid chlorides. Reacting aldonolactones with hydroxylamine in aqueous solution leads quantitatively to aldonohydroxamic acids (14). There are two tautomers of hydroxamic acid.

The preparation of micro/nano spheres and capsules, consisting of mixtures of a polysaccharide such as starch and a protein is patented (15).The particles contain a cross-linked outer layer (interfacial cross-linking) with hydroxamic acid groups on the outside. In the patent examples based on combinations of chitosan and collagen are described. Other proteins and polysaccharides are also claimed. The use of a mixture of a protein and a polysaccharide is supposed to improve the stability of the particles against enzymes and the treatment with hydroxylamine. The introduced hydroxamic acid groups form complexes with different metal ions. The products are used in pharmaceutical and cosmetic applications.

Lutfor et al. (16) also reacted polymethylacrylate sago starch graft polymers with hydroxylamine in aqueous methanol (methanol:water 5:1) at 72 °C during two hours. Products containing 30 % sago starch showed maximal swelling in water. The decrease of the swelling in aqueous NaCl, $CaCl_2$ and $FeCl_3$ is explained by the complexing of the multivalent cations by the hydroxamic acid ligands.

The partial hydroxyethylation of carboxymethyl starch with ethylene oxide results in the hydroxyethylesters of carboxymethyl ethers (17). The conversion with hydroxylamine should lead to hydroxamic acid derivatives.

References

1. BP 808,290 (1959) Monsanto.
2. Modified starches, Properties and Uses. CRC Press. 1987; 188.
3. JP 04247062 (1992).
4. JP 60162899 (1985).
5. JP 6220505 (1987).
6. JP 6245794 (1987).
7. EO 200830 (1986) Kyoritsu Yuko Co.
8. Czechoslovak patent 165918.
9. M.R. Lutfor et al., European Polymer J. 2000; 36(10): 2105-2113.
10. Verraest et al., Carbohydr. Pol. 1998; 37: 209-214.
11. B.W. Zhang et al., React. Pol. 1993; 20(3): 207-216.
12. EP 846,495 (1998).
13. M.L. Hassam et al., J. Appl. Pol. Sci. 2003; 87(4): 666-670.
14. L. Salmon et al., Carbohydr. Res. 2001; 335: 195.
15. USP 6,132,750 (2000).
16. M.R. Lutfor et al., Carbohydr. Pol. 2001; 45: 95-100.
17. USP 3,092,619 (1963).

Ester migration in carbohydrates

Under certain conditions partial esters of polyhydroxy compounds like carbohydrates are showing ester migration. An ester group moves from one OH group to an other. Fisher (1) described the shift of partial acetylated glycerol esters. This shift is intramolecular via cyclic orthocarboxylic acid derivatives (2) as was demonstrated with ^{14}C labelling. The ester migration is catalyzed by bases.

Thevenet et al. (3) studied, in the course of the preparation of surfactants from sucrose, the monoacetylation of sucrose with acid chlorides such as octanoylchloride, allyl- and octylchloroformiate in aqueous alkaline solution at pH10. The reaction was mainly determined by the competition between esterification, migration and saponification. During the esterification the C_2-OH is the most reactive group, but in the end, caused by migration of the ester groups there is a substantial substitution of the primary OH groups. If sucrose mono esters are hydrolyzed in pure water a fast reduction of the C_2-OH ester is observed and an increase in the C_3-OH ester, next the C_6OH ester is formed from the C_3-OH. The esters on the primary position are more stable against hydrolysis than the esters on the secondary position. At the start of the hydrolysis a di-substituted sucrose was found. The probable explanation is a transesterification between two mono substituted sucrose molecules resulting in a sucrose- and di-substituted sucrose (inter-molecular ester migration). In a 50% solution the ester is more stable, but also under this condition several migrations occur. If sucrose is esterified in a concentrated aqueous solution there will be polysubstitution caused by hydrophobic clustering. If a co-solvent such as tetrahydrofuran is used this hydrophobic aggregation will not occur.

Cabral et al. (4) studied the esterification of 5,6-isopropylidene-ascorbic acid (IAA).

IAA

By protecting the C_5-OH and the C_6-OH group the study was simplified. If IAA is esterified in aqueous alkaline solution there is C_2-OH substitution. In the absence of proton donors there is C_3-OH substitution. The C_3-OH is the most acid OH group with a pK_a =4. The result of the esterification was also influenced by the degree in which ester migration took place. Under certain conditions a migration of the ester groups from C_3-OH to the C_2-OH is occurring.

Jarowenko (5) described the industrial acetylation of granular starch in aqueous slurry with both acetic acid anhydride as well as vinylacetate. The described products have a low DS. The reaction is carried out at ambient temperature under slightly alkaline conditions. Although analogous to alkali catalyzed etherification often C_2-OH substitution is assumed (kinetic control) there is no hard evidence published. The high reactivity of the C_2-OH is related to the relatively acid character of this group with a pK_a=12.4. The fast acetylation of potato starch with acetic acid anhydride results into heterogeneously substituted products with mainly surface substitution (6).

The dry retorting of starch with acetic acid results in starch acetates with predominantly C_6-OH substitution (thermodynamic control).

There is hardly any literature published about the distribution of acetyl groups in acetylated starch over the starch granules and molecules and the hydroxyl groups of the glucose units. The influence of this distribution on retrogradation and enzymatic degradation has certainly been observed (7). This calls for a detailed research into these esterifications with attention for the occurrence of intra- and inter molecular ester migration besides esterification and hydrolysis. Ester migration might also be the result of an after-treatment of derivatives produced in aqueous slurry resulting in more homogeneous substitution.

References

1. E. Fisher, Ber. 1920; 53: 1621.

2. A.P. Doerschutz, J. Am. Chem. Soc. 1952; 74: 4202.

3. S. Thévenet et al., Carbohydr. Res. 1999; 318: 52-66.

4. J. Cabral et al., J. Carbohydr. Chem. 1998; 17(9): 1321-1329.

5. W. Jarowenko, Modified Starches, Properties and Uses. CRC Press. 1987; 59.

6. P.A.M. Steeneken et al., Starch 96. Carbohydrate Research Foundation. 1997; 47.

7. Personal communication.

Unusual carbohydrate esters, ferulic-, gallic- and lactic acid esters

Ferulic acid esters

A typical chemical characteristic of pectins is the presence of acetyl- and ferulic acid (4-hydroxy-3-methoxycinnamic acid) ester groups.

ferulic acid ester

The ferulic acid ester groups are bound to neutral sugar chains, about 30% is bound to the arabinan chains and the remainder to the galactan chains. The ferulic acid groups can cause cross-linking resulting in a viscosity increase or gelling of aqueous pectin solutions. The cross-linking is achieved by oxidation with peroxidase/H_2O_2 or with ammonium persulfate. Oxidative cross-linking is an alternative for gel formation in an acid/sugar system or by Ca complexing. The ferulic acid groups of the arabinan side chains are predominantly involved in the gel formation. The conditions for the extraction of the pectins are important. Too drastic treatment with hot acid leads to decomposition of the instable arabinosyl bonds and the liberation of ferulic acid which has to be oxidized before it can contribute to gelatinization (1). This phenomenon has been extensively studied by Oosterveld et al. (2).

Branan ferulate is a polysaccharide extracted with aqueous alkali from corn bran. The M_w weight was determined with GPC and is about 200.000 Dalton. Based on methylation and enzymatic degradation it was concluded that the polysaccharide backbone is a high substituted (1→4)-β-D-xylan. The central carbohydrate chains are substituted with α-L-arabinofuranoside and α-D-glucopyranosyluronic acid residues. Some of the arabinose side-chains are substituted with ferulic acid. Amino acid analysis showed the presence of 4.4%w/w amino acid (0.22% hydroxyproline). Cross-linking with peroxidase whereby two ferulic acid groups are coupled results in a three dimensional structure. The structure is shown in

In the presence of water this product forms a hydrogel which can be used as a wound management aid (3).

In the case of bread improvers ferulic acid esters and peroxidase possibly also play a role. The probable reaction might be presented as

In vitro this reaction has been demonstrated (4).

Novo Nordisk claims in a patent (5) *"a method for causing gelling or increase of viscosity of an aqueous medium containing a gellable polymeric material having substituents with fenolic hydroxyl groups, wherein an effective amount of laccase is added to said aqueous medium"*.

Laccase is a copper containing polyfenoloxidase. With oxygen and in the presence of laccase cross-linking occurs. Cross-linking increases the viscosity and gelling can occur. As substrates polysaccharides from wheat and sugar beet like arabinoxylanes, heteroxylanes and pectins are used, they contain small amounts of ferulic acid esters. The cross-linking can also be carried out with peroxidases and H_2O_2 or with persulfate. Such products are used in various applications ranging from feed to food and pharmacy.

Novo Nordisk (5) suggests the esterification of carbohydrate based polymers such as starch, cellulose etc. with acids like ferulic acid with ferulic esterase. Esters and ethers with phenolic substituents based on starch and cellulose are hardly described. As far as the authors are aware, a cross-linking of starch (derivatives) with enzymes has not been reported.

Hyaluronate is an important component of the synovial fluid which acts as a lubricant of joints. During an inflammation the molecular weight decreases and the lubricating effect becomes less, resulting in pain. This decrease in molecular weight is caused by OH radicals. To prevent this decrease Hercules (6) patented the use of derivatives of polysaccharides with anti-oxidant substituents. These derivatives are prepared by reacting polysaccharides with acid groups (carboxymethyl starch is mentioned) with suitable reagents. The derivatization of hyaluronic acid with 3,5-dibutyl-4-hydroxybenzoyl chloride in dry polar aprotic solvents such as N-methylpyrrolidine is described. To dissolve the salts of these polysaccharides in these polar solvents they are converted into the tetraalkylammonium salts. After the reaction, the sodium salts are prepared by ion-exchange. The reaction can be combined with cross-linking for example with pyromellitic acid anhydride. The optimum DS is about 0.0025. The products are applied for the treatment of arthritis, as drug delivery vehicle, to improve the healing of chronic wounds and in cosmetic formulations. It seems obvious to prepare the ferulic acid esters in the same way.

Ferulic acid has been reported to have many physiological functions, including antioxidant, antimicrobial, anti-inflammatory, anti-thrombosis and anti-cancer activities. It also protects against coronary disease, lowers cholesterol and increases sperm viability. Because of these properties and its low toxicity, ferulic acid is now widely used in the food and cosmetic industries (7).

Ferulic acid in its free form hardly reaches the large intestine. To obtain this effect, starch derivatives were made. Shiji Ou et al. (8) described the preparation of starch ferulates with ferulic acid chloride in DMSO with sulphuric acid as catalyst at 100 °C and 40 minutes. The product had a DS 0.036. It appeared that degradation by diastase is small and the micro-organisms in the large intestine set the ferulic acid free even faster and better than from dietary fibres (wheat bran).

The reaction of starch with ferulic acid chloride or with ferulic acid thiosulphate in aqueous alkaline slurry is proposed. Application of a starch based carrier for other acids than ferulic acid to the large intestine offers possibilities.

Gallic acid esters

Gallic acid (3,4,5 trihydroxybenzoic acid) is an important component of tannins (9). Tannins are extracted from plants, especially from sumac leaves. Tannins are applied in the leather industry, as mordant, in inks and pharmacy. Tannins are complex compounds and can be divided into two groups, those which hydrolyse under influence of acids and enzymes and those which condensate during this treatment into complex insoluble polymeric compounds. "Turkish tannin", a glucose derivative esterified at the $C_{1,2,4}$ and C_6-OH with gallic acid or its oligomers is an example of a hydrolysable tannin.

Rhubarb contains glucogallan-β-glucose monogallate - which decomposes upon hydrolysis into one molecule glucose and one molecule gallic acid (10). Fischer already demonstrated that esterification of glucose with gallic acid resulted into a compound similar to tannin.

Because of the antioxidant properties gallic acid containing compounds are presently receiving attention from the food industry. An example of such a compound is epi-gallocatechin-3 gallate, a component of green tea.

epi gallocatechine - 3 - gallate:

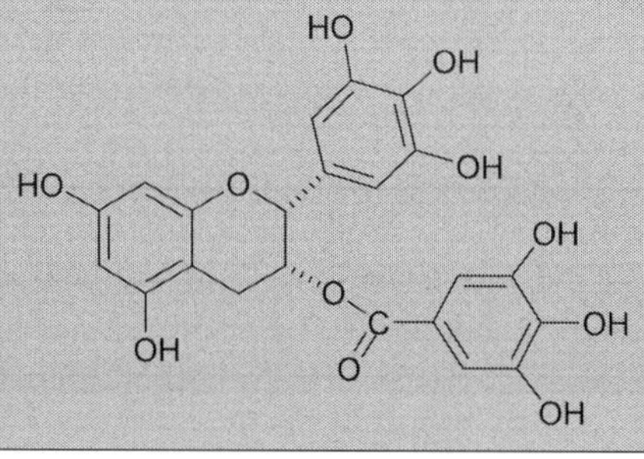

This compound is supposed to be responsible for the health promoting properties of green tea. The effect is supposed to originate in the interaction of polyphenols with urokinase. Also the inhibition of the tyrosinase kinase receptor phosphorylation has been mentioned. It is a radical scavenger and as such it has a positive influence on inflammations connected with rheumatism (11, 12, 13).

Phenols react easily with oxidizing radicals

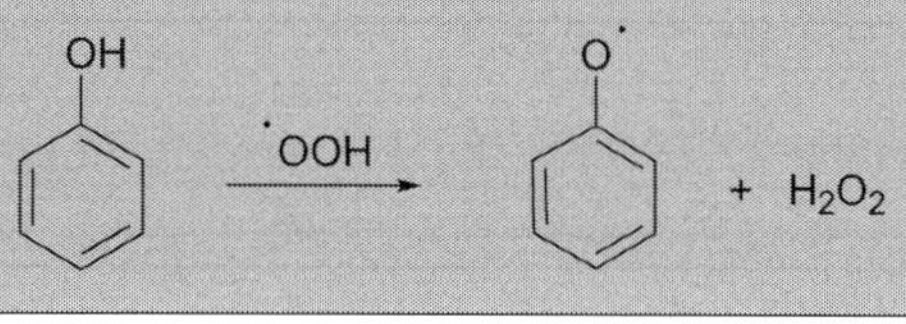

The formed phenoxy radicals are relatively stable and can react further into quinones or react with each other. Propylgallate is industrially used as an anti-oxidant. It can also prevent the thermal degradation of polysaccharides and starch (14, 15). In this degradation oxygen, metal ions and radicals play a role. Mixtures of propylgallate and sulfite (1:3) can prevent loss of viscosity of galactomannan during heating; it has been shown that such mixtures also reduce the decomposition of starch. Care has to be taken with this treatment as small amounts of bisulfite enhance the degradation of starch. Propylgallate can be replaced by ferulic acid and to a certain extend by L-ascorbic acid.

Potier et al. (16) has described the synthesis of a series of gallic acid esters of saccharose. Sucrose and two equivalents of DMPA dissolved in DMF-triethanolamine (2:1) was added to a solution of 8 equivalents of 3,4,5-tribenzylgalloyl chloride in CH_2Cl at 0 °C and stirred overnight at ambient temperature. Sucrose gallate-8(benzyl) (SG-8) was formed with a yield of 51%.

By hydrogenolysis (Pd/C) the benzyl groups were removed. Purification over a C_{18} silica column resulted in a SG-8 as a pink powder. By varying the amount of reagent products with different content of gallic acid were obtained:2,3,6 or 8 groups per sucrose unit. The anti-oxidant properties were related to the number of gallic acid groups indicating no saturation by steric hindrance at higher DS. The anti-oxidant properties were measured with diphenyl picryl hydrazyl (DPPH) and showed to be better than the natural gallotannins because of the higher gallic acid content.

Enzymes can also be used in the synthesis of such products (17). The preparation of antitumor phenolic acid sugar esters is patented.

Phenolic acid alcohol ester + sugar $\xrightarrow{\text{esterase}}$ phenolic acid sugar ester.

Xiaowei Yu et al. (18) studied the enzymatic synthesis of gallic acid esters from gallic acid and alcohols in organic solvents with microencapsulated tannase from Aspergillus niger. The microencapsulated enzyme showed higher synthetic activity than the free enzyme. The highest yield was found in benzene and the most suitable substrate 1-propanol.

Lixian and Seib (19) have described the preparation of an antioxidant based on ascorbic acid and gallic acid. By reacting ascorbic acid with gallic acid in the presence of sulfuric acid at 25°-30° C during 26 to 44 hours a mixture of gallic acid esters is formed. The yield of $C_5 + C_6$ esters (2:5) is 90%. From this mixture the C_6 ester is separated by fractionated crystallization with a yield of 40%.The stability of the 6-gallate at pH 5 and 25°C is limited because of hydrolysis. Also ester migration from C_6 to C_5 occurs.

L-ascorbyl-6-gallate:

The esters of 1 and 2 were compared with the ethers of 3 and 4

The ethers were prepared by reacting the 6-bromium derivatives of ascorbic acid with 3 and 4 in alkaline environment. The antioxidant activity was measured in methyllinoleate and found to be 1>2=3>4. The 6-gallate is less effective than propylgallate but better than L-ascorbyl-6-palmitate and L-ascorbic acid.

The preparation of starch or starch derivative based gallates offers possibilities. Esterification in concentrated sulphuric acid probably will result into low molecular esters, a synthesis with enzymes (20) into high molecular derivatives.

Lactic acid esters

Lactic acid esters of alcohols are used as fragrant, solvent, cosmetic ingredient etc. Ethyllactate is generally regarded as a non-toxic solvent. When preparing lactic acid esters the presence of the OH group in the lactic acid leads easily to the formation of a lactide.

Via the lactide polylactic acid is prepared. Polylactic acid (See above) is a thermoplastic material, is biodegradable and is used in bio-medical applications.

From et al. (21) have described the esterification of lactic acid in alcohols with enzymes (Novozyme SP435 from Candida antartica). With butanol the highest reaction velocity was observed. With higher alcohols lower reaction velocities were observed possibly due to the solvatation in the solvent. The synthesis of butyl lactate was carried out in a two phase system at 70 °C. The alcohol was dissolved in hexane with ample lactic acid added. After the reaction the enzyme particles were filtered off, the aqueous phase removed and the solvent distilled. The product was isolated by distillation at decreased pressure (70 °C, 25 mm Hg). The reaction velocities of the secondary alcohols were appreciably lower than of the primary alcohols. No dimerization or oligomerization was observed. The used lipase catalyzes both enantiomers in about the same way.

Bousquet et al. (22) described the enzymatic synthesis of lactic acid esters of carbohydrates. α-butylglucoside lactate was prepared by transesterification of α-butylglucoside with butyl-lactate using the enzyme Novozyme 435 (lipase adsorbed on an acrylic resin).

alfa - butylglucoside lactate

After a reaction time of 40 hours at 50 °C a yield of 67% was obtained. By removing the butanol from the reaction mixture the yield could be increased to 95%. By H and C^{13}-NMR it was shown that only C_6-OH substitution occurred. These products can be used in cosmetic formulations because of their hydratating properties.

A variation on this synthesis is the preparation of n-octyl-β-D-glucopyranoside lactate using the enzyme Novozyme 435 by the transesterification of n-octyl-β-D-glucopyranoside and ethyl acetate in acetone (23). After a reaction time of 12 hours at 60 °C a molar yield of 87% n-octyl-β-D-glucopyranoside lactate is obtained at an overall conversion of 90%. Molecular sieves are used to remove the ethylalcohol and to shift the equilibrium to ester formation. Acetone is used to increase the solubility of the solid precursor. This solvent easily evaporates and is permitted for use in food.

Both syntheses have the advantage over the chemical derivatization of region selective substitution without incorporating oligomeric lactic acid chains. The derivatization of other sugars with other enzymes f.i proteases seems feasible.

Wolf Walsrode patented the synthesis of cellulose based lactic acid esters (24), in particular the hydroxy alkyl cellulose-2-hydroxycarboxylic acid ester. Hydroxyethyl- and hydroxypropyl cellulose ether dissolved in DMSO, N-methylpyrolidone, dimethylacetamide etc. are heated during 5 hours at 130 °C with L-lactide. The reaction mixture is put into acetone and the product flocculates. After washing it is dried. Products with a MS-lactate of 0.6-1.0-1.5 and 1.8 are described.

Although the reaction products of hydroxyethyl cellulose with glycolide are claimed only the reaction products with lactide are described. One example describes the reaction with an oligomeric lactic acid. The hydroxyethyl- and hydroxypropyl cellulose ethers used have a DS of 0.68 or 0.92. The produced lactate esters have a MS-total of >1.5. The biological degradation of these products is good contrary to other non-ionic cellulose ethers. The solubility is temperature dependant, only below a certain temperature the products dissolve, above this temperature flocculation occurs. If the MS-lactate is >3 the flocculation temperature is below room temperature. This flocculation behaviour resembles the properties of methyl cellulose and is related to the presence of hydrophobic substituents and their effect on the structure of the surrounding water molecules.

A Japanese patent (25) describes the reaction of lactic acid with starch.
"A mixture of 20 g L-lactic acid and 10 g a-starch were gelatinized during 30 min at 130 °C and subsequently cooled in a nitrogen atmosphere to 80 °C. Next the product was heated under vacuum for 8 - 16 hours at temperatures from 80° to 120 °C".
The formed lactic acid esters were thermoplastic and biodegradable.

Sikora et al. (26) used a microwave when reacting starch with α-amino- and α-hydroxy acids. Using an extruder for this type of process looks feasible as starch succinates can also be produced in this way (27).

National Starch patented (28) the preparation of lactide modified starch derivatives. According to the patent it is possible to react starch in aqueous alkaline slurry with lactide. A cooled high amylose corn starch in alkaline slurry is reacted with powdered lactide at pH 7.5. The pH is kept constant during the reaction by adding a 3% NaOH solution. During the reaction the 6 ring structure of the lactide is ruptured and a di-lactic acid ester is formed. The efficiency of the reaction is reported to be low. The products are used for the preparation of starch foams by extrusion. They are very flexible, strong and well compatible with synthetic polymers. Also starch derivatives such as hydroxyethyl starch can be used as starting material.

To improve the properties of polylactic acid Ohya et al. (29) prepared graft polymers by reacting L-lactide with trimethylsilylpullulan in an organic solvent such as THF with t-BuOH as catalyst. The trimethylsilylation is necessary to dissolve the pullulan in organic solvents. The introduction of hydrophilic segments and the branched structure improved the biodegradation. Li et al (30) published a similar study based on dextran.

In the preparation of lactic acid esters it is important to take the chirality into account. In the human body R-lactic acid is slowly metabolized while L-lactic acid is formed in large amounts and converted with lactatedehydrogenase.

The (industrial) research into the possibilities of lactic acid esters of carbohydrates, in particular starch, has hardly started. It is possible to use lactic acid, oligomeric lactic acid and lactides. The industrial preparation of starch lactates preferably will have to be carried out in aqueous slurry, aqueous alcohol or via a dry processing. There are several products known based on heated mixtures of polylactic acid and starches, this to prepare improved bioplastics. Starch-polylactic acid graft copolymers are maybe a better option. For this purpose starch derivatives are a good choice especially the low DS hydroxyethyl starch as the number of primary OH-groups is higher than in starch. If these primary hydroxyl groups react selectively in esterification then hydroxyethyl starch is more reactive than native starch combinations with hydrolysis, oxidation, cross-linking widen the scope. Graft polymers of polylactic acid and cyclodextrins offer the possibility of star shaped polymers. The preparation of polyesters of glucosides is maybe an alternative for the preparation of polyurethane foams, alkyd resins, surfactants, etc.

References

1. F. Guillon et al., Carbohydr. Polym. 1990; 12: 353-371.
2. A. Oosterveld et al., Carbohydr. Res. 2000; 329: 199-207.
3. L.L. Loyd et al., Carbohydr. Polym. 1998; 37: 315-322.
4. Chemisch Magazine. 1995; June: 264.
5. WO 97/27221 (1997) Novo Nordisk.
6. EP 0749982 (1996) Hercules.
7. Shiyi Ou et al., J. Sci. Food Agric. 2004; 84(11): 1261-1269.
8. Shiyi Ou et al., Food Chemistry. 2001; 74(1): 91-95.
9. A.F. Holleman, Leerboek der Organische Chemie. Wolters. 1955: 587.
10. A.E. de Groot, J.F. Engbersen, Bioorganische Chemie. Wageningen Acad. Press. 1992: 514.
11. J. Jankum et al., Nature. 1997; 387: 561.

12. Chemisch Weekblad. 1999; 2.

13. Welt am Sonntag. 1999; April.

14. S.E. Hill et al., J. Sci., Food Agric. 1999; 79: 471-475.

15. P. Sriburi et al., Food Hydrocolloids 1999; 13: 177-183.

16. P. Potier et al., Tetrahedron Letters. 1999; 40: 3387-3390.

17. JP 09322794 (1998).

18. Xiaowei Yu et al., J. Mol. Cat. B. 2004; 30(2): 69-73.

19. L. Gan, P.A. Seib, J. Carbohydr. Chem. 1998; 17(3): 397-404.

20. WO 9603440 (1996) Novo Nordisk.

21. M. From et al., Biotechnology Letters. 1997; 19(4): 315.

22. M.P. Bousquet et al., Biotechnol. Bioeng. 1999; 62(2): 225.

23. C. Torres et al., Biotechnology Letters. 2000; 22: 331-334.

24. DE 19731575 (1999).

25. JP 0892313 (1996).

26. M. Sikora et al., Polish J. Food & Nutr. Sci. 1997; 6(20): 23-30.

27. L. Wang, Starch/Stärke. 1997; 3: 116.

28. EP 1,178,054 (2001) National Starch.

29. Y. Ohya et al., Macromolecules. 1998; 31: 4662-4665.

30. Y.X. Li et al., Polymer. 1998; 39(14): 3087.

Super absorbents based on starch

Super absorbents, named Super Slurper for its ability to absorb large amounts of water, were first introduced by the U.S. Department of Agriculture in 1976 (1). The patent covered starch-polyacrilonitrile graft copolymers, alkaline saponified. The carboxylic groups are neutralized with NaOH to pH 6-7.

starch-polyacrylonitrile graftpolymer

The water absorption is about 1000 g/g, absorption of aqueous salt solutions is far less and about 40 g/g.

The currently used products are mainly based on the sodium salts of cross-linked polyacrilic acid and used as hygienic pads for the absorption of body fluids. The inside of the pads are covered with a perforated water repellent film. The fluids can pass this film and are absorbed by a layer of compressed cellulose (fluff). From this it is transported to an inner layer of super absorbent where it is retained even under pressure.

The driving force for the moisture absorption is osmosis. Because of the hydratation of the cations the negative charge of the anions is less compensated resulting in a repulsion of the negatively charged carboxylic groups. The polymer chains are stretched and the super absorbent swells. A hydrogel is formed, elastic and easily deformed (2, 3).

Super absorbents based on polyacrylates are not well biodegradable. Mixing or grafting with biodegradable polysaccharides improves the biodegradability However at a starch content of over 25% the absorption properties markedly decrease. Mixing with dextrins improves the absorption synergistically (4). Adding a polycaprolactone and titanylsulfate to a mixture of polysaccharide and polyacrylate and cross-linking with Al-compounds is claimed to result in a super absorbent while improving biodegradability (5).

Much effort has been put into the development of super absorbents based on carboxymethyl starch (CMS) and carboxymethyl cellulose (CMC). By optimizing the DS, degree of cross-linking, particle size etc. products were obtained which could compete with the polyacrilic acid based products. If in the synthesis of starch based super absorbents a small amount of the acrylonitrile is replaced by a monomer with a sulfonic group the absorption improves (6).

Lin et al. (7) described the synthesis of starch sulfate based polyacrylonitrile graft copolymers. The starch sulfation was carried out with starch dissolved in DMAc/LiCl with DMSO/SO$_3$ complex. The starch sulfates were reacted with acrylonitrile and hydrolyzed. The effect of the DS-sulfate and the degree of grafting (PAN/AGU) on the water and salt solution absorption (g/g) are shown in Table 6.

Table 6. Effect of DS-sulfate and grafting on absorption.			
DS sulfate	g-PAN/AGU	Water absorption	Salt solution absorption
0	213.0	820	61.5
0.35	250.9	1442	114.4
0.81	290.6	1510	126.4
1.14	229.0	1616	122.2
1.71	217.0	1750	113.4

From the results it can be concluded that introducing the sulfate group increases the absorption. At a DS of 0.35 the major part of the effect is already reached. The effects of a lower DS are as yet not known. For industrial purposes this synthesis is not suitable because of the use of an exotic solvent and reagent. Perhaps a change to the easier introduced sulfoalkyl groups might be the answer. The following reagents for the preparation of sulfoalkyl ethers are proposed:

$$\text{H}_2\text{C}-\overset{\text{H}_2}{\text{C}}-\overset{\text{O}}{\underset{\text{O}}{\text{S}}}\overset{\text{O}}{\diagdown}\qquad \text{(1,3-propane sultone)}$$

$$\text{Cl}-\overset{\text{H}_2}{\text{C}}-\overset{\text{H}_2}{\text{C}}-\overset{\text{H}_2}{\text{C}}-\overset{\overset{\text{O}}{\|}}{\underset{\|}{\text{S}}}{\underset{\text{O}}{}}-\text{O}-\text{Na}$$

$$\text{Cl}-\overset{\text{H}_2}{\text{C}}-\overset{\text{H}}{\underset{\text{OH}}{\text{C}}}-\overset{\text{H}_2}{\text{C}}-\overset{\overset{\text{O}}{\|}}{\underset{\underset{\text{O}}{\|}}{\text{S}}}-\text{O}-\text{Na}$$

$$\text{H}_2\text{C}=\overset{}{\underset{\text{H}}{\text{C}}}-\overset{\overset{\text{O}}{\|}}{\underset{\underset{\text{O}}{\|}}{\text{S}}}-\text{O}-\text{Na}$$

$$\text{Cl}-\overset{\text{H}_2}{\text{C}}-\overset{\overset{\text{O}}{\|}}{\underset{\underset{\text{O}}{\|}}{\text{S}}}-\text{O}-\text{Na} \quad \text{en} \quad \text{Cl}-\overset{\text{H}_2}{\text{C}}-\overset{\text{H}_2}{\text{C}}-\overset{\overset{\text{O}}{\|}}{\underset{\underset{\text{O}}{\|}}{\text{S}}}-\text{O}-\text{Na}$$

Chloromethanesulfonate can be easily prepared by reacting methylene chloride with sodium sulfite. The sulfomethylation of starch with sodium chloromethanesulfonate is hardly described and can be carried out analogous to the carboxymethylation. Sulfomethylation of cellulose is already known for a long time (8, 9).

Combining carboxymethylation and sulfoalkylation with cross-linking of starch, by preference in a one step synthesis, could result into a more effective super absorbent. Starting with cross-linked dialdehyde starch reacted with sodium bisulfite super absorbents with sulfonate groups can be prepared (10). A more or less similar method is the reaction of maleic acid half esters of starch with sodium bisulfite and mercapto compounds (11). Maleic acid esters of starch as such can also be used (12).

Besides the above mentioned type of products also starch phosphates can be used as super absorbent (13). Relatively high DS starch phosphates are prepared by heating starch with phosphoric acid or their salts in the presence of urea. Besides phosphate groups also carbamate groups are introduced. Such products are claimed to be hydrolysis resistant and well biological degradable.

A Japanese patent (14) describes a variation in the way of cross-linking. Carboxymethyl starch is heated with amino acids like aspartic acid and glutamic acid. Cross-linking occurs by reaction of the OH groups of the starch with the amino acids. A Korean patent (15) describes a similar reaction by heating starch with adipic acid and maleic acid.

Seidel et al. (16) described the preparation of hydrogels by heating a carboxymethyl starch with a DS 0.45 with organic acids such as malic acid, citric acid, malonic acid, maleic acid, glutaric acid, adipic acid and tartaric acid. The different cross-linkers had a large influence on the strength of the hydrogels (4%). This difference is explained by the difference in length of the spacer and the number of carboxylic groups in the acid. Citric acid gave the best network.

The absorption of hydrogels based on cross-linked carboxymethyl cellulose and hydroxyethyl cellulose can be increased by incorporating spacers into the cross-linker. Reaction of polyethylene glycols with two molecules divinylsulfone results in such a cross-linker. By varying the molecular weight in the polyethylene glycol different spacer lengths are obtained (17). Starch derivatives certainly will show the same effect.

The cross-linking of carboxymethyl starch by reacting the carboxyl groups with its own OH groups in the starch is already known for a long time (18, 19, 20). Starch in aqueous alkaline solution is carboxymethylated and after isolation and drying heated for some time at 90 °C. Another method combines the carboxymethylation and cross-linking by carrying out the reaction in an extruder and using a slight under dosage of alkali with respect to the monochloroacetate.

Surface cross-linking is also used as a method to improve the absorption capacity (21, 22). A mixture of acrylic acid and sodium acrylate is polymerized in aqueous solution with a radical initiator and a poly unsaturated cross-linker in the presence of a blowing agent (23). The resulting microcellular hydrogel is milled, dried and subsequently surface cross-linked in an organic solvent. The blowing agent such as a carbonate creates the open structure. The surface cross-linking increases the absorption under pressure and gel strength after absorption.

Starch gels with an open structure can be obtained by starting with micro-porous starch. This starch is obtained by degrading granular starch with α-amylase or glucoamylase (24). Cross-linking of these porous starch particles gives a higher mechanical strength to the particles. To increase the absorption the micro-porous particles can be treated with surface modifying agents. To increase the lipophilicity compounds such as methylcellulose, poly-N-vinyl-2-pyrrolidoen etc. are adsorbed on the surface. The same effect is achieved by covering the surface of the starch particle with fatty acid groups (reacting the starch surface with octenyl succinic acid anhydride). Such products can be used as carrier for fragrances, vitamins, pesticides etc. Surface hydrophobation can also be achieved by treating the microporous starch particles with polysiloxanes (25). The products are used to absorb bad odor compounds like thiols and ammonia from aqueous solutions or from the gas phase as present in hygienic pads. Incorporation of cyclodextrins, either via a covalent or ionic bond or mechanically mixed, in super absorbents also results in odor absorbing products (26).

To improve the biodegradability of a super absorbent a preparation based on a low substituted starch derivative has been patented (27). Starch or a low DS starch derivative is "dissolved" in water and dispersed in oil. The droplets of the starch solution are cross-linked and filtered off. After swelling of the particles in water during 48 hours the particles are isolated by filtration and washed with alcohol. The products have an absorption capacity of 25g/g of which 91% is achieved in the first 5 min.

Decreasing the particle size in a super absorbent increases the absorption velocity, however also gel blocking is occurring. The small gel particles coagulate thereby impeding flow. Agglomeration of the small particles can prevent this phenomenon (28). An other possibility is to coat the particles with very fine inorganic oxide particles (29). The trend towards still smaller hygienic pads has resulted into using more super absorbent and less fluff cellulose. This also leads to gel blocking. An additional cross-linking at the end of the process with multivalent metal ions (Al^{3+}) can prevent this (30).

Encapsulating a non-colloidal organic or inorganic solid filler (starch can be used) with a hydrogel forming coating can result into a high performance super absorbent (31). Non colloidal particles are individually dispersed in a non water miscible solvent in the presence of a surfactant. An ethylenic unsaturated monomer able to polymerize to a hydrogel forming polymer is suspended in the non water miscible solvent together with an initiator. Next the filler particles are coated with the hydrogel forming polymer.

Graft copolymerization of acrylamide, potato starch and an ultra fine mineral such as kaolin followed by an alkaline hydrolysis leads to a super absorbent with high water absorption. Products with $-CONH_2$, $-COONa$ and $-COOH$ groups are superior to those products containing only one of these groups separately (32).

Super absorbents can not only be used in hygienic pads but also for moisturizing soil, coating of seeds, in medical applications for wound dressing and body powders. Wet books can be dried with a super absorbent (33). In conclusion the application of super absorbents in the preparation of lyocell (cellulose) fibres with improved absorption properties can be mentioned (34).

References

1. Starch, Chemistry and Technology. Academic Press. 1984; 406.
2. Chem. 2 Weekblad. 2001; 97(11): 12.
3. Chem. Mag. 1995; 9: 370.
4. USP 4,483,050.
5. DE 4206857 A1 (1993) Stockhausen.
6. Modified Starches, Properties and Uses. CRC Press. 1987; 156.
7. Doo-Won Lim et al., J. Appl. Polym. Sci. 2001; 79: 1423-1430.
8. USP 2,820,788 (1958).
9. USP 2,891,057 (1959).
10. EP 920874 (1999).
11. USP 6,063,914 (2000) Stockhausen.
12. EP 0714914 A1 Degussa.
13. DE 19859123 C1 (2000).
14. EP 0637594 (1994) Nippon Shokubai Co.
15. CA 133 209569.
16. C. Seidel, Starch/Stärke. 2001; 53: 305-310.
17. A. Sanino et al., J. Appl. Polym. Sci. 2003; 90: 168-174.
18. EP 0538904 A2 (1992) Kimberley-Clark Co.

19. USP 5,079,354 (1992).

20. NOA 9100249 (1991) AVEBE.

21. USP 4,587,308.

22. USP 4,507,438.

23. EP 0644207 (1994) Nalco.

24. WO 89/04842 (1989).

25. USP 6,147,028 (2002) P&G.

26. WO 9964485 Stockhausen.

27. EP 0900807 (1997) ATO.

28. USP 5,180,622.

29. USP 3,932,322.

30. USP 6,433,058 (2002).

31. WO 97/27884 (1997).

32. J.Wu et al., Polymer. 2003; 44(21): 6513-6520.

33. The Alchemist. The Chem Web Magazine. 8/10/2003.

34. K.Y.Lim et al., European Polymer J. 2003; 39(11): 2115-2120.

The application of enzymes involved in starch derivatization: Some recent developments

It has been since years that enzymes for the conversion of starch into products like dextrose, maltodextrins, glucose and maltose syrups are applied on an industrial scale. For the manufacture of these products hydrolases, such as α- and β-amylases, isoamylases, pullulanases and glucoamylases, are used. Specific transferases like phosphorylase and branching enzyme are mainly restricted to laboratory research for the preparation of α-D-glucose-1-phosphate, chain elongated and branched starch; cyclodextrin-glucosyltransferase (CGTase) however is currently used on an industrial scale for the production of cyclodextrins (CD's). Isomerases are applied industrially for the isomerisation of glucose to fructose.

Although the traditional products of enzymatically converted starch and sugars find their way nowadays in the food sector mainly, in the future also more and more technical applications will emerge. This is the result of applying new enzymes and unusual substrates, reaction conditions or media. Below some recent developments are summarized.

Application of hydrolases

Glucoamylase, isoamylase and CGT-ase are enzymes with a starch binding domain in their polymeric structure, separated from the active sites. These enzymes first attach to the substrate before product formation takes place. It has been shown that these enzymes can enter the pores of the starch granule and then bind to the starch molecule. Through the action of glucoamylase the pores are excavated while glucose is formed. The breakdown of granular starch by glucoamylase is dependent on the starch type; not only the pores but also other structural characteristics play a role. In this respect waxy maize granules break down very well due to the large number of non-reducing end-groups in the starch molecule. This is important because glucoamylase attacks amylose and amylopectin from the non-reducing end of the molecules. It is interesting to note that starch granules can be broken down in the presence of limiting amounts of water at 37 °C with the formation of 10 - 50 % of glucose within the granule. Hardly any reversion could be detected. Waxy maize starch granules with more than 20 % glucose taste sweet! (1).

As mentioned glucoamylases have a starch binding domain separated from their active sites. By the action of 1 - 3 % of glucoamylase on granular starch in water at pH 3.5 - 6 and with a temperature of 40 - 55 °C during a short period of time (0.1 - 15 minutes) the enzyme binds to the granular surface without substantial hydrolysis of the starch. By subsequently decreasing the pH to 2.0 in 5 minutes the enzyme will be inactivated; after that the pH can be brought to 6.0 again. This results in a starch with a granular surface with hydrophobic properties. This product is better suited for applications that ask for a higher affinity for fats and oils, as in cosmetics (2).

The reactivity of granular starch during hydroxyethylation can be increased by applying enzymes like glucoamylases and α- and β-amylases. Due to this activation higher substituted products can be obtained without concomitant swelling. This treatment should be an alternative for the use of large quantities of swelling inhibitor or organic solvents. However, the scientific explanation of the observed effects is not clear (3).

Rice starch can be made porous by hydrolysis of the granular starch with a mixture of glucoamylase and α-amylase. It was found that the derivative can absorb substances such as water, oils, ethanol, and gaseous odors/aromas. The absorption capacity of porous starch was saturated for coffee oil when starch and coffee oil were brought in contact with each other for 30 minute at 70 °C; the absorbed coffee oil can be released completely by heating during 10 minute at 130 °C. Compared to native starch the porous starch releases the absorbed substance less easy. Porous starch releases only 80% of the absorbed coffee oil during 2 weeks at room temperature (4). A suspension of native starch or cross-linked native granular starch in water is treated with debranching enzymes at temperatures that are a few degrees below the initial swelling temperature. A pre-treatment with maltogenic α-amylase may be executed. The resulting products can be obtained by simple filtration and may be applied in foodstuffs and pharmaceutical products. It has the potential to substitute gelatin, a.o.

Some years ago the enzymatic synthesis of 1-O-benzyl-α-gucoside (BG) and 1-O-benzyl-α-maltoside (BM) by reaction of α-amylases with starch and benzylalcohol in water has been described (5). The reaction was carried out at 40 °C at varying pH for 3 days. Of the enzymes tested α-amylase from Aspergillus oryzae gave the best results. Optimal synthesis of BG was at pH 5.5 and of BM at pH 8.0, whereas at pH 5.0 BM was hydrolysed very quickly to glucose and BG by the α–amylase; at pH 8.0 BG did not react anyhow. Under no conditions BG could be hydrolysed. For comparison, maltotriose was not converted neither at pH 5.5 nor at pH 8.0. The optimal D.P. for glucoside formation still has to be determined. In the referred literature it is stated that: "*Although the enzymatic synthesis of various glycosides from monosaccharides or oligosaccharides have been reported, the direct synthesis of glycosides from starch has never been reported*". This statement seems not correct. After all, in 1994 Takahisa Nishimura and coworkers published about the glycosylation of hydrochinon with a specifically isolated α-amylase while "soluble starch" was the initial substrate. In this reaction more than 30% of the quinone was converted into 4-hydroxyphenyl-O-α-D-glucopyranoside. The glucosylation of phenolic compounds was intended in particular to increase the water solubility, Today these phenolic compounds are seen as interesting molecules for their antioxidant properties (6).

Shin et al (7) described the synthesis of hydroxybenzyl-α-glucosides by transglucosylation of soluble starch and hydroxybenzylalcohols. They applied amyloglucosidases from Rhizopus sp. The optimal glucoside formation is at 45 °C and pH 5.0.

In general β-galactosidases are used for the hydrolysis of lactose. In addition, some recent publications mention other possibilities of this type of enzymes. See the following example of a sugar conversion:

maltopentaose + lactose $\xrightarrow[\text{40°C, 5 u}]{\text{β-galactosidase}}$ oligosaccharides

(Gal 1-4 Glc 1-(4Glc)n4Glc)

It may be wondered whether the relatively short side chains of amylopectin molecules can be extended with galactose groups similarly (8).

Another striking feature of β-galactosidase is the galactosidation of hydroxyethylcellulose. In a first orientation the galactosidation reaction takes place in water in particular at varying pH and temperatures. In this cited literature the properties of the products are described summarily. Mainly the OH-groups of the hydroxyethyl moiety play a role in the galactosidation reaction. The reaction efficiency is measured by monitoring the amount of hydrogen peroxide formed by the oxidation of the galactose derivatives with galactose oxidase. The mentioned derivatisation of hydroxyethylcellulose was also described in the patent literature (9, 10). It is obvious that starch derivatives have potential in analogous reactions.

It is known that lipase can perform esterification of low molecular sugars with fatty acids. The synthesis of esters from maltooligosaccharides with a D.P. >2 with lipases in organic solvents is laborious due to the decreasing solubility of the sugars. It was shown that a 6'-O-capronic acid ester of maltose, produced in t-butanol with lipase, could be subjected to transglucosidation in water with CD, maltodextrins or starch as donor molecules. This resulted in maltooligosaccharide esters. By using H and ^{13}C-NMR and MALDI_TOF MS 6''-O-caproyl maltotriose and 6'''-O-caproyl maltotetraose could be identified (11). Almost nothing has been published on similar reactions with polysaccharides. Therefore a publication is referred to, which describes the synthesis of esters from fatty acids with cellulose acetates (12). The applied cellulose acetates are soluble in acetonitrile and with an immobilised lipase (NOVOZYME-Candida antarctica) 35 % of the added fatty acid could be bound to the cellulose under optimal conditions.

In this context also the enzymatic conversion of solubilized amylose with subtilisin in an organic solvent may be mentioned:

$$R-OH \ + \ H_3C-(CH_2)_8-\underset{\underset{O}{\|}}{C}-O-\underset{H}{C}=CH_2 \ \xrightarrow{\text{subtilisin/isooctane}} \ H_3C-(CH_2)_8-\underset{\underset{O}{\|}}{C}-O-R$$

The enzyme preparation is soluble in isooctane. The reaction was performed with both amylose films and with amylose powders (13).

Hydroxyethyl cellulose and cationactive guar were made hydrophobic by protease and lipase action with vinylstearate in N,N-dimethylacetamide and t-butylmethylether (14). Also starch can be made hydrophobic when it reacts with hydrophobic diketene derivatives in organic solvents as DMSO and DMAc by action of hydrolases, such as lipase. The reaction is carried out at 50 °C for several hours. The reaction mixture is brought into isopropylalcohol and the resulting precipitate was washed (15).

Another typical enzymatic synthesis is the production of vinylglucose esters with a protease form Streptococcus sp. The conversion of glucose with divinyladipate takes place in DMF at temperatures between 30 °C and 60 °C. After 24 hours reaction a yield of 90% was obtained. Above 60 °C inactivation takes place and at 80 °C the region-selectivity decreases. From research with ^{13}C-NMR it was concluded that C_6-substitution had occurred (16). In a similar way the divinyladipic acid esters of sucrose and trehalose could be synthesized with commercially available proteases and lipases. Then these di-esters were used as monomers in enzymatic poly-condensation reactions with different diols under the formation of linear poly-esters (17). These esterifications are just first orientations. Further research may be focussed at practical processes and applicable products.

Application of transferases

The last few years several interesting application opportunities have been described about transferases like phosphorylases, branching enzymes and CGT-ases. A nice example is the application of phosphorylase, catalysing the synthesis of 2-deoxymalto-oligosaccharides (18). With a primer and phosphate D-glucal can be converted into 2-deoxy-α-D-glucopyranosylphosphate with phosphorylase.

$$\text{D-glucal} \;+\; P_i \;\xrightarrow{(Glc)_n}\; \text{2-deoxy-α-D-glucosylphosphate}$$

The minimal chain length of the primer must be four units. In fact a two step reaction was involved. First the primer chain is extended with a deoxy glucose unit, as follows:

$$\text{D-glucal} \;+\; (Glc)_n \;\xrightarrow{[P_i]}\; \text{2-dGlc-}(Glc)_n$$

Next phosphorylysis of the deoxy sugar takes place with the formation of 2-deoxy-α-D-glucose phosphate under restitution of the primer. Instead of maltotetraose also soluble starch could be used as primer. This only works with an excess of glucal in order to prevent starch phosphorylysis. Further, the reaction can be directed by varying the amount of phosphate. In the presence of equimolar quantities of phosphate and glucan mainly 2-deoxy-α-D-glucosylphosphate is yielded (to 62%). At a limiting phosphate concentration (0.05 eq) 2-deoxy-maltooligosaccharides are obtained with a yield of about 45%. 2-deoxy-maltooligosaccharides with an average chain length of 20 proved to be insoluble in water, but were well soluble in water diluted alkali. Products with a DP of 12 were still water soluble. The 2-deoxy-oligosaccharides are broken down by both α-amylase and glucoamylase. According to the authors of the article the 2-hydroxy group may not be of importance for the binding to the enzymes mentioned. As a consequence 6-deoxy-D-glucal and D-xylal cannot be converted. It looks very interesting to study the properties of starch with chains that are elongated by 2-deoxy-glucose groups. Moreover the synthesis of glucal may be industrially interesting.

Several other typical reactions with phosphorylase have been described.

α-D-glucose-1-phosphate can be enzymatically polymerised with phosphorylase in the presence of polytetrahydrofuran. The conversion yields an amylose-polyTHF (polymer-polymer) inclusion complex (19).

Another interesting reaction is the synthesis of new disaccharides with phosphorylases. As an example, maltose can be converted into β-D-glucose-1-phosphate and glucose by maltose phosphorylase in the presence of inorganic phosphate. The formed glucose can successively be removed by fermentation and the remaining β-D-glucose-1-phosphate can react with another sugar like fucose into a new sugar (20).

Also the synthesis of 5 new oligosaccharides by glucosyl transfer from β-D-glucose-1-phosphate to isokestose and nystose is possible with Thermo-anaerobacter brockii kojibiose phosphorylase (21).

Other transferases such as dextran sucrase and alternan sucrase from Leuconostoc mesenteroides can glucosylate α-butyl- and α-octyl-glucopyranoside, with sucrose as the glucosyl donor. This is the first description of an enzymatic glucosylation of α-alkyl-glucosides with more than two carbon atoms in the alkyl moiety. This may result in new possibilities for the modification of molecules in the APG group of surfactants (22).

D-enzyme (disproportionating enzyme), which is a 4-α-glucanotransferase, was first discovered by Peat in the potato tuber in 1956. The enzyme catalyses the transfer of maltooligosaccharides (usually maltose) from the one 1,4-α-D-glucan molecule to the other or to glucose (disproportionation). Maltooligosaccharides are effective donor molecules but also short chain amylose and amylopectin can act as donor. The similar D-enzyme from bacterial origin is called amylomaltase. The role of D-enzyme in the biosynthesis of starch is still subject of discussion (23). It is an interesting finding that D-enzyme is not only able to catalyse the disproportionation reaction but also cyclisation reactions. In this way cycloamylose could be obtained in high yield (> 95%) by converting "synthetic" amylose (i.e. side chain free amylose). Cycloamylose has no reducing capacity, is well soluble in water, does not retrograde and formes complexes with several compounds and has many application possibilities (24). This D-enzyme catalysed cyclisation of amylose was also shown with amylopectin as substrate (25). In this respect it is important to note that the conversion of starch in water solution with glucosyltransferases (a.o. D-enzyme) has been patented by AVEBE (26). Glucosyltransferases belong to the class of enzymes with the formal nomenclature of α-1,4-glucosyltransferases, which comprise both thermostable and thermolabile enzymes. Examples are the thermostable glucosyltransferase from Thermus thermophilus and the labile D-enzyme from potatojuice. The thermostable enzymes can react with a starch solution at 70 °C and pH 6.5 for a few hours, the D-enzyme works at 30 °C for 48 hours. The resulting products have a high molecular weight but are low in viscosity; moreover these products show thermoreversible gel formation! The patent which describes these products mentions the broad application potential in food, pharma and cosmetic products and in adhesives and drilling muds. At present the mechanism of the reaction of D-enzyme with different substrates is not yet clearly understood. Meanwhile it has been shown that also the branching enzyme catalyses the production of cyclic structures from amylose and amylopectin (27).

The reaction potential of the transferase CGT-ase is complex. The best-known reaction is the formation of cyclodextrins (G6,-7,-8). Also the reversed reaction, which is a coupling reaction, can take place. Further, "disproportionation" ($G_m + G_n \leftrightarrow G_{m-y} + G_{n+y}$) is possible. It has been shown that from within the starch granule cyclodextrins can be released with CGT-ase. At 37 °C during 20 hours a maximum of 1.4% CD's was obtained. When combined with isoamylase the amount of product increased 2.6 times. From waxy maize a maximum of 3.4% was obtained (28).

Many publications describe the application of CGT-ase for the preparation of oligosaccharides and glycosides (transfer reactions). In this way glycerol can be converted with CGT-ase and starch as donor into a series of glucosides with maltooligosyl residues which are bound to glycerol (29). Another interesting phenomenon is the conversion of 1,5-anhydro-D-fructose (GAF) with β-cyclodextrin, catalysed by CGT-ase. The resulting products have different length of the malto-oligosyl residues - bound to the C-3 of GAF. These products can be converted with glucoamylase into glucosyl-1,5-anhydro-fructose (30). And finally the enzyme CGT-ase is also able to form cyclic structures from amylopectin that deviate from the usual α-, β- and γ-dextrins (31).

The cycloalternan-forming enzyme (CAFE) is able to convert alternan into cycloalternan. A partially purified enzyme preparation which consists of two protein molecules can also use malto-oligosaccharides to produce cycloalternan. The first step in the reaction is the disproportionation of a D-glucopyranosyl moiety of the non-reducing end of one malto-oligosaccharide to the non-reducing end of an other malto-oligosaccharide, under the formation of an isomaltosyl unit at the non-reducing end. Next, CAFE transfers the isomaltosyl part of the non-reducing end of another isomaltosyl oligosaccharide, followed by the cyclization to cycloalternan (32).

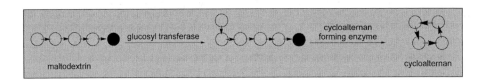

Interesting properties of the amylosucrase enzyme are described in a few patents of AVENTIS (33). Amylosucrase catalyses the conversion of sucrose into linear polyglucans.

$$sucrose \xrightarrow{\text{amylosucrase}} \alpha\text{-D-1,4 glucan}$$

Water soluble amyloses and branched products ("dextrins") can also serve as substrate for chain elongation. Amylosucrases can be obtained /isolated from bacteria (Neisseria polysaccharae sp.). The concerted action of amylosucrase and other enzymes, such as the branching enzyme may give interesting products. Apparently the amylosucrase produced insoluble polysaccharides can be applied at the preparation of microspheres, resistant starch and as filler in tablets. Chain elongation of starch with amylosucrase and sucrose may be an interesting alternative for the application of phosphorylase and α-glucose-1-phosphate.

Application of lyases

In two publications (34, 35) the reaction of starch with α-1,4-glucan lyase is described.

$$\text{starch (n)} \xrightarrow{\text{glucan lyase}} \text{1,5-anhydro-D-fructose + starch (n-1)}$$

The enzyme belongs to the sub-class of polysaccharide lyases. It has been developed by Danisco and is stable for at least 30 days at room temperature. With maltose as substrate 1,5 anhydro-fructose and glucose are formed in equimolar quantities. Amylopectin and glycogen give 1,5 anhydro-fructose and limit-dextrins. From starch the compound is formed with a yield of 50%. When in addition pullulanase is applied the yield may increase to 90%. 1,5 Anhydro-fructose might be applied as low calorie sweetener. One can think of several other possibilities!

In a "one pot" synthesis of sugars from glycerol and an aldehyde one of the steps is the conversion of dihydroxyacetone phosphate with butanol with fructose-1,6-biphosphate aldolase (Staphylococcus carnosis) (36).

Application of oxidoreductases

Research on the potential possibilities of several oxidizing enzymes for the derivatisation of starch and derivatives has increased substantially. Within the scope of this review it is worthwhile to refer to the results of two publications.

The first article is titled: "The Cetus process revisited: a novel enzymatic alternative for the production of aldose free D-fructose" (37). In the Cetus process glucose is converted into glucosone with pyranose-2-oxidase; glucosone is further hydrogenised to fructose catalytically (38). In the new process glucose is also oxidized enzymatically, but now with a pyranose-2-oxidase from Trametus multicolor. The enzyme activity is stabilized by albumin and catalase. Catalase degrades the hydrogen peroxide formed. Glucosone was obtained with a yield of 98%. This glucosone could be reduced quantitatively to fructose by an NAD(P) dependent reductase from Candida tenuis. In this way two different enzymes were successfully applied for continuous cofactor regeneration. Through this simple method D-fructose could be prepared deliberately free from glucose. From an industrial point of view a process where the enzymatic production of glucose is highly integrated with glucosone production may be very relevant.

The enzymatic oxidation of oligosaccharides to oligosaccharides with a gluconic acid end group has been described several years ago (39). A novel carbohydrate oxidase from Microdochium strains is able to oxidise the glucose end groups in maltodextrins and cellodextrins with higher activity than the monomer glucose. These enzymes may be applied in bread baking (40). It is interesting to investigate the potential production of similar oligosaccharides with glucosone end groups.

Quite extensive research is done on the oxidation of sugars and also starch with TEMPO/chlorine. The chlorine (eventually with bromium as catalyst) helps to convert TEMPO in a reactive molecule able to oxidise carbohydrates. It fits within the scope of this chapter that three publications refer to the use of an oxidising enzyme in stead of co-oxidants like chlorine et al. (41, 42, 43). The NOVO patent (41) is very broad: "Modification of polysaccharides by means of a phenol oxidising enzyme". When the phenol oxidase is a peroxidase also hydrogen peroxide is needed for the reaction together with necessary "enhancing agents". In the patent many enzymes and enhancers are mentioned; the only described example is the oxidation of cellulose with laccase/TEMPO in water.

The Finnish patent (42) focuses on the oxidation of starch with laccase/TEMPO. Example 2 describes the oxidation of in water suspended granular starch with oxygen at pH 5 for 3 hours. The enzymatic reaction catalyses the introduction of carbonyl- and carboxyl groups. The reaction with cellulose is patented as well (43). The reaction scheme is as follows:

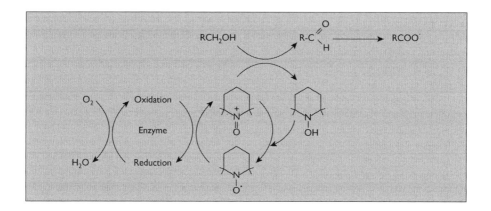

A typical manganese oxidase has been patented (44). The enzyme catalyses the oxidation of a benzyl alcohol derivative (a.o.) with oxygen in the presence of manganese cations as mediator. The methylol-groups are converted into aldehyde groups. Mn^{2+}-ions are oxidised to Mn^{3+}-ions, which further oxidise a suitable substrate. The reduced form of the enzyme is oxidised by oxygen. The relevance of this type of enzyme systems in the oxidation of carbohydrates is obvious.

Application of isomerases

Cross-linked crystalline enzyme preparations (xylose isomerase), when trapped in a "packed bed reactor", can be applied for the slow isomerisation of L-arabinose in L-ribulose and the epimerisation into L-ribose. By varying the reaction conditions the quantities of effluent sugars could be changed: arabinose (100% - 0%), ribulose (0% - 55%) and ribose (0% - 100%). Using the same procedure cross-linked crystalline xylanase preparations catalyse the synthesis of xylose and xylobiose from xylotetraose (45). In the Netherlands the research group of Sheldon found an alternative process for the preparation of cross-linked enzyme crystals. They managed precipitation of the enzymes in aggregates. Successively these aggregates are cross-linked with the usual glutaraldehyde. (46).

In summary, in literature many interesting developments in enzymatic conversions of starch and its derivatives, using hydrolases, transferases, lyases, oxidoreductases and isomerases link to new possibilities for carbohydrate derivatisation with potential for future industrial application.

References

1. Yeon-Kye Kim, John F. Robyt, Carbohydr. Res. 1999; 318: 129 - 34.
2. WO 0224938 (2002).
3. USP 605 1700 (18/04/2000) Grain Processing Corp.
4. Yao Weirong et al., Starch/Stärke. 2002; 54: 269 - 263.
5. Jong Yi Park et al., Biotechnology Letters. 1999; 21: 81 - 86.
6. Takahisha Nishimura et al., J. Ferm. Bioeng. 1994; 78(1): 31 - 36.
7. Hwa Kyoung Shin et al., Biotechnology Letters. 2000; 22: 321 - 325.
8. JP 04279596 (05/10/1992).
9. Jun Li et al., Carbohydr. Res. 1999; 316: 133 - 37.
10. WO 9849201 (05/11/1998) Hercules.
11. Peter Degn et al., Carbohydr. Res. 2000; 329: 57 - 6334.
12. V. Sereti et al., J. Biotechnol. 1998; 55: 219 - 223.
13. Ferdinanda F. Bruno et al., Macromolecules. 1995 ; 28: 8881 - 8883.
14. Qu-Ming Gu, Polymer Preprints. 2000; 41(2): 1834.
15. USPA 20020123624 (2002) Hercules.
16. Masaru Kitigawa et al., Biotechnology Letters. 2000; 22: 879 - 882.
17. Oh-Yin Park et al., Biotechnol. Bioeng. 2000; 70 (2): 208.
18. Britte Evers et al., Starch/Stärke. 1995; 47: 434.
19. Jun-ichi Kadokawa et al., Chem. Commun. 2001; 5: 449 - 450
20. Kazuo Aisaka et al., J. Biosci. Bioeng. 2003; 90 (2): 208 - 213.
21. Hideka Okada et al., Carbohydr. Res. 2003; 338: 879 - 885.
22. Gaëtan Richard et al., Carbohydr. Res. 2003; 338: 855 - 864.
23. Takeshi Takaha et al., J. Biol. Chem. 1993; 268 (2):1391 - 1396
24. Takeshi Takaha et al., J. Biol. Chem. 1996; 271 (6): 2902 - 2908
25. Eur. P. 0675 137 (03/04/1995).
26. WO 98/15347 (16/04/1998) AVEBE.
27. Hiroki Takata et al., Carbohydr. Res. 1996; 295: 91 - 101.
28. Yeon-Kye Kim, John F. Robyt, Carbohydr. Res. 2000; 328: 509 - 515.
29. Irofumi Nakano et al., J. Biosci. Bioeng. 2003; 95 (6): 583 - 588.
30. Kazuhiro Yoshinaga et al., Carbohydr. Res. 2003; 38 (21): 2221 - 2225.
31. Eur. P. 0710674 (13/09/1995).
32. Yeon-Kye Kim et al., Carbohydr. Res. 2003; 338 (21): 2213 - 2220.
33. WO 9967412 (29/12/1999) Aventis R&T GMBH, WO 0012589 (09/03/2000) Aventis R&T GMBH, WO 0022155 (20/04/2000) Aventis R&T GMBH.
34. Carbohydrates in Europe. 1999; 26: 9.
35. Søren M. Anderson et al., Carbohydr. Res. 2002; 337: 873 - 890.
36. Schoevaart R. et al., Chem. Commun. 1999; 24: 2465 -2466.

37. Christian Leitner et al., Biocatalysis and Biotransformation. 1998; 16: 365 - 382.

38. Eur. P. 88103 (1981).

39. Lin Shuen Fuh et al., Biochem. Biophys. Acta. 1991; 1118 (1): 41 - 47.

40. USP 6165761 (26/12/2000) Novo Nordisk.

41. WO 99/32652 Novo Nordisk.

42. WO 99/23240 Valton Teknillinen Tutkimuskeskus - Finland.

43. WO 99/23117.

44. Patentschrift DE 19909546 (29/06/2000).

45. Matti Leisola et al., Biotechnol. Bioeng. 2001; 72 (4): 501 - 506.

46. Chemisch Weekblad. 2001; 2 (22 September).

The application of peroxidases

General

Peroxidases are oxidoreductases that use hydrogen peroxide or alkyl-hydroperoxides as acceptor. Peroxidases are found in animals, plants and microorganisms. They catalyse the conversion of a wide variety of substrates (1, 2, 3).

The so called "classical peroxidase reaction" refers to the enzyme catalysed conversion of organic compounds, thereby using hydrogen peroxide as an oxidant.

In classical peroxidases ferri protoporphyrin IX is the prosthetic group and imidazole the fifth iron ligand. The characteristic activity of peroxidases is a one-electron oxidation; for details of the mechanism see the cited literature.

Haloperoxidases are peroxidases that incorporate halogen atoms in organic molecules. The reaction needs the presence of hydrogen peroxide.

$$SH + H_2O_2 + H^{\oplus} + X^{\ominus} \longrightarrow SX + 2H_2O$$

SH is an organic substrate, X is chloride, bromide or iodine, SX is the halogenated product. Little is known about the natural function of haloperoxidases.

Haloperoxidases always introduce halogen atoms in an electron dense part of the substrate thereby substituting a proton or adding a halogen and a hydroxyl group to a double or triple bond. The products that are formed are fully identical to the products formed with chemical halogenation agents under the same conditions. In literature it is supposed that all bromation reactions occur via $HOBr/Br_2$ and that all vanadium-haloperoxidases also form HOX. The non-selective enzymes may be important in synthesis reactions because HOX is formed relatively slow and controllable! Non-haem haloperoxidases are superior towards haem containing haloperoxidases due to the greater steadiness under denaturating conditions.

Chloroperoxidases catalyse not only the reactions typical for peroxidases but also those typical for catalases and monooxygenases and they further catalyse halogenation reactions in the presence of halogenide (except fluoride) and H_2O_2. Interestingly sometimes a pseudo halogenide, like thiocyanate can take the role of the halogenide (lactoperoxidase). The chloroperoxidase of the fungus Caldariomyces fumago is presently the far most studied enzyme; it is commercially available (Sigma). The immobilisation of the enzyme on aminopropyl controlled Pore Glass is described. The disadvantage of the enzyme is the low optimal pH 3.0 at halogenation. The enzyme has a relatively high catalase activity. This means that hydrogen peroxide has to be added slowly and in portions. The enzyme has a labile haem group as also applicable to lactoperoxidase and horseradish peroxidases have.

The reactions that are catalysed by peroxidases as a group can be categorised as follows:

oxidative dehydrogenation

$$2SH + H_2O_2 \longrightarrow 2S + 2H_2O$$

oxidative halogenation

$$SH + H_2O_2 + \overset{\oplus}{H} + \overset{\ominus}{X} \longrightarrow SX + 2H_2O$$

H_2O_2 dismutation

$$2H_2O_2 \longrightarrow 2H_2O + O_2$$

oxygen transfer reaction

$$SH + H_2O_2 \longrightarrow SOH + H_2O$$

More specifically reactions are involved like oxidation of alcohols and phenols, conversion of alkenes in halohydrins, epoxidation of alkenes, conversion of aromatic amines into nitroso compounds and of sulphides into sulphoxides.

Much scientific research has been performed with the enzyme horseradish peroxidase and the previously mentioned chloroperoxidase from Caldariomyces fumago. The potential of other enzymes such as for instance soybean peroxidase are under study.

The last few years the interest in the applications of peroxidases is increasing. In this respect the following communications and citations are illustrative.

Cross-linking of pectins by peroxidase action

Apart from the presence of acetyl groups, pectins have another striking chemical feature in the presence of ferula acid ester groups bound to neutral sugar chains. 30% is bound to arabinan chains and the rest to galactan chains. The ferula acid groups may give rise to crosslink reactions, resulting in a viscosity increase or gelation by peroxidase/H_2O_2 or ammonium persulphate action. Oxidative cross-linking is an alternative for gel formation in an acid/sugar system or by calcium complexation. The ferula acid groups which are bound to the arabinan side chains are mostly involved in the gelling process. Important are the extraction conditions: too drastic treatment with a hot acid solution results in the decay of the labile arabinosyl bonds and in the release of ferula acid that has to be oxidised first before gelation can take place. Much research in this area has been done in particular by the University of Wageningen in the Netherlands (4, 5).

Enzymatic cross-linking of proteins

Modification of the native protein structure usually leads to major changes in the functionality of the protein. This may offer interesting opportunities for industrial application, for example because the protein has greater stability at higher temperatures in water or in organic solvents. A nice example is the enzymatic cross-linking of proteins. Advantages of enzymatic cross-linking are the mild reaction conditions and the specificity and controllability of the process. Moreover starter reagents and end products are less toxic than when chemical treatment is applied (6).

The former ATO (currently A&F Wageningen) has carried out research on the peroxidase catalysed cross-linking of proteins. In this system protein dimers, tetramers and polymers are obtained with better foam forming, gelling and film forming properties. This opens doors to interesting industrial applications.

Peroxidases for the improvement of bread

In bread making gluten from wheat flour are very important for the quality of the bread. Wheat dough consists mainly of gluten (gliadin and glutenin), lipids, starch and other non-starch carbohydrates. Several different processes take part in the development of a dough. First of all during the kneading the structure of the protein complex that is formed after mixing flour and water is disrupted. By the kneading the protein chains are stretched and slide along each other. During the leavening of the dough these proteins form a hugh network, the gluten. A good aggregation of the proteins in the gluten is important for the gas retaining ability of the dough and eventually determines the volume and firmness of the dough. Nowadays peroxidases are applied for the improvement of bread making. Up till now two model systems exist for the strengthening of the dough by peroxidases. It was shown in vitro that two ferula acid molecules (bound to arabinoxylan) can be coupled oxidatively by peroxidase activity and H_2O_2. The reaction scheme is given below (7).

This coupling reaction gives a carbohydrate gel that is very well water retaining and hence results in a firm bread dough. The second model system describes the aggregation of proteins with carbohydrates, also via *ferulic* acid. Until recently peroxidases were obtained exclusively from horseradish. However the application of this peroxidase in bread is illegal. DSM has discovered another vegetable source of peroxidase. These peroxidases have market potential.

Removing phenol with horseradish

Horseradish has yet other application possibilities than just culinary. Jean Marc Bollage and Jerry Dec, both affiliated to the Penn State Centre for Bioremediation & Detoxification showed in their research that a ground horseradish preparation can clean up industrial waste water that is polluted with phenol. Phenol is a constituent of waste water from steel production, ore mines, paper bleaching and other industrial processes. According to the authors horseradish removes 95% of the phenol for a price which is only half the price of conventional chemical and filtration techniques. Horseradish contains the enzyme peroxidase that is added to the filthy water together with peroxide, thereby converting the phenols into insoluble polymers. These polymers can be filtered from the solution. The ground horseradish can be reused a 30 times. The researchers have applied for a patent. They haven't yet found a method to get rid of the spent horseradish (8).

Improve your washing with fungal enzymes

Chemists of the University of Amsterdam, the Netherlands developed a fungal enzyme that can help to manufacture less damaging detergents. Washing powders already contain these proteinaceous compounds that have the advantage of being well degradable. This new enzyme produces the same bleaching compounds that are produced from relatively harmful and badly degradable chemicals, used in the present generation of detergents. The research started years ago and after some time marine algae were discovered containing an enzyme that was able to produce bleaching compounds from seawater with bromium. The detergents industry showed interest but preferred the chlorine bleaching agents because these are less dangerous. Through scientific literature the research group also traced a fungus that produces an enzyme with the same bleaching activity but now with chlorine. The group succeeded in obtaining a sufficiently purified enzyme preparation from their fungus, which is a Curvularia species. In principle the enzyme with the name chloroperoxidase might be added to washing powder. It produces hypochlorite with chloride which is also a constituent of the powder.

Hypochlorite has a strong bleaching activity. The use of the enzyme has the advantage of omitting relatively harmful bleaching agents, like perborates, from the washing powder. However it is not yet possible to produce the enzyme on an industrial scale; only milligrams per liter could be achieved instead of grams. It is the aim to increase the production of the enzyme by transferring the gene that is responsible for the enzyme synthesis to an other fungal species. The fungus Curvularia that produces the enzyme by nature has a regulatory inhibition to prevent overproduction. This inhibition is probably absent in other, non-related species; this may increase the production of the enzyme substantially. After constructing a genetically improved fungus, it can be applied on an industrial scale. The produced enzyme can be added to the detergents. The active site of the enzyme contains vanadium. Therefore a small quantity of vanadium is added to the washing powder (9).

Lignin peroxidase and bio-pulping

For the preparation of a good paper quality from wood pulp the lignin has to be eliminated, eventually followed by bleaching. A set of chemical methods exists to realise this. Chloride and chlorodioxide have the disadvantage of AOX formation. The combination of oxygen and H_2O_2 may be an alternative; H_2O_2 is hardly effective. The application of microbial cultures or enzymes may be a different approach (bio-pulping). This method aims at the decrease in the use of energy and chemicals. The degradation of lignin by microorganisms like "white rot fungi" will eventually be realised with lignin peroxidase. Probably also other redox systems play a role in the final mineralisation. Lignin peroxidase converts accessible aromatic centres into instable cation-active radicals through a one-electron oxidation. The radicals may further undergo several non-enzymatic reactions. Model systems showed cleavage reactions in the aromatic ring, being the formation of quinones and aromatic acids. It is a disadvantage that the spontaneous coupling reactions of the instable radicals result in "restructuring" rather than degradation. Soy bean peroxidase has also potential to remove lignin from wood pulp. The soy enzyme is said to be cost effective, thermally and chemically stable. It can be extracted from soy hulls and is stable over a pH range of 1.5 - 13 and till above 70 °C. The soy enzyme has a higher redox potential than the horseradish enzyme (10, 11).

The application of haloperoxidases for the preparation of glucosone

Haloperoxidases have an interesting application in the conversion of propylene into propylene chlorohydrin with hydrogen peroxide that is released in the oxidation of glucose to glucoson (by glucose-2-oxidase). The propylene chlorohydrin can be converted with a halohydrinperoxidase into propylene oxide. This process has been developed in the eighties by the biotech company Cetus. Their aim was not only using the main product, but also the H_2O_2 produced. Normally H_2O_2 is degraded by catalase. D-glucoson may be converted into fructose or 2-ketogluconate. Recently the method showed a revival through the application in a modified process (12, 13, 14, 15).

Oxidation of alcohols with chloroperoxidases

Contrary to other peroxidases that in general are restricted to the oxidation of phenols through free radical intermediates, the chloroperoxidase of Caldariomyces fumago (CPO) is able to convert allyl-, propargyl or benzalcohols onto aldehydes in the presence of H_2O_2. However, reactive aldehydes such as 5-hydroxymethylfurfural are partly further oxidised. For instance one of the two aldehyde groups may be converted into a carboxylic group (16).

The chloroperoxidase from Caldariomyces fumago is a haem peroxidase which contains iron (III) protoporphyrin (IX) as prosthetic group. The extracellular enzyme can be isolated in applicable quantities. CPO catalyses a wide variety of oxidations with the aid of H_2O_2 through the synthesis of an oxo (V) porphyrin intermediate without using expensive cofactors.

Derivatisation of starch with peroxidases

As shown in the previous paragraphs much research is reported on peroxidases. Obviously it is legitimate to discuss in addition the potential of peroxidases in the derivatisation of starch and its derivatives. It was already described earlier that haloperoxidases play a role in the preparation of glucoson. The liberated H_2O_2 is used for the synthesis of propylene chlorohydrin (and further to propylene oxide) from propene. In the derivatisation of starch chlorohydrins are applied as reagents in ether forming reactions such as cationisation. It might be feasible to apply peroxidases for the synthesis of these reagents, possibly in the presence of starch and an immediately following etherification (17, 18, 19).

Unsaturated starch derivatives might be converted into derivatives with chlorohydrin substituents (which may be converted further into epoxy groups). A characteristic conversion is the polymerisation of phenolic compounds with peroxidases. In a previous chapter the possibilities were mentioned of cross-linking/gelation of starch derivatives with phenolic substituents, using peroxidases.

Grafting of gallic acid esters on chitosan with peroxidase has been described recently. A similar graft polymerisation on starch might also very well be possible. Several other monomers seem interesting!

Starch in an aqueous solution can be oxidised by peroxidase (from horseradish) in the presence of TEMPO and hydrogen peroxide to C_6-aldehyde starch at room temperature and at pH 5. (Partial) conversion of this product with Girard's reagent (2 hours and at 40 °C) results in a cationic derivative which may be useful as wet strength additive.

Finally it may be pointed out that the enzymatic synthesis of oxidative reagents like hypochlorite and peracids have potential. They may be applied whether in situ or not in the preparation of oxidised starches.

In conclusion: as just shown the first applications of peroxidases in starch derivatisation are described. The evident versatility of peroxidase activity on different substrates may predict an increased interest of peroxidases in starch (bio)chemistry for industrial application.

References

1. Stefano Colonna et al., TIBTECH. 1999; 17: 163.
2. M.C.R. Franssen, Biocatalysis. 1994; 10: 87 - 111.
3. US 5391488 (1995).
4. Fabienne Guillon et al., Carbohydr. Polym. 1990; 12: 353 - 71.
5. A. Oosterveld et al., Carbohydr. Res. 2000; 328: 199 - 207.
6. Chemisch Weekblad. 2002; 2(16 February): 14.

7. Chem. Mag. 1995; June: 264.

8. ATO informatie, 1995.

9. De Volkskrant. 12/09/1992.

10. T.Gliese et al., Wochenblatt für Papierfabrikation. 1994; 310

11. EP 0598538 (1993).

12. US 4267641 Cetus.

13. EP 004221 (1981) Cetus.

14. US 4351902 (1981) Cetus.

15. Christian Leitner et al., Biocatalysis and Biotransformation. 1998; 16: 365 - 82.

16. M.P.J. van Deurzen et al., J. Carbohydr. Chem. 1997; 16(3): 299 - 309.

17. Laurant Vachoud et al., Enzyme and Microbial Technology. 2001; 29: 380 - 85.

18. WO 96/06909 (1996) Degussa.

19. WO 0183887 (08/11/2001).

Some interesting oligosaccharides

Raffinose

The sugar raffinose, O-α-D-galactopyranosyl-(1-6)-O-α-D-glucopyranosyl-(1-2)-β-D-fructofuranoside, is also called melitose or melitriose.

Raffinose is found in sugar beet molasses, cotton seed and various other plants. It is probably the most occurring oligosaccharide. Raffinose is converted into meliobiose and fructose with invertase or diluted acid. With emulsinase saccharose and galactose are formed.

Raffinose has no reducing properties; it crystallizes with 5 mol water and has no sweet taste. During the crystallization of saccharose the raffinose concentration increases in the mother liquor. At a raffinose concentration, calculated on sugar, of more than 1% the crystallization of saccharose becomes difficult and different crystal shapes (more needles) are formed. An average sugar factory can produce 5-10 tonnes raffinose per day. Raffinose is converted into saccharose and galactose with α-galactosidase, the galactose is removed during the usual lime treatment. With recombinant DNA techniques it has become possible to produce pure α-galactosidase, an advantage over the old types as they are contaminated with invertase. Invertase converts saccharose into glucose and fructose (1). Instead of removal an effective production of raffinose can be chosen (2). It is possible to remove or isolate the raffinose more effectively using membranes (3, 4). Also an improved method for the crystallization of raffinose has been patented (5).

Raffinose can also be produced enzymatically. A Japanese patent describes the production of raffinose from α-galactosyl fructoside and sucrose with α-galactosidase from Pycronporus cinabarinus 1F06139. Alpha-galactosylfructoside is prepared from saccharose and lactose with levansucrase (6). Alpha-glucosidase has also been used to reduce the raffinose content in the meal of pulses. Raffinose causes flatulence after the eating of pulses. In India research was carried out into the use of a fungal extract as a source of α-glucosidase. This extract is a by-product of the gibberellic acid production and is available almost for free. Treatment of the meal of pulses resulted in almost complete degradation of the raffinose (7). The effect of α-rays on the content of oligosaccharides was also investigated. Radiation of mung beans resulted in a decrease of the oligosaccharide content of 70-80% in comparison of a decrease of 30% of untreated beans. No relation with flatulence was observed.

Raffinose can be used for the following applications:
- diet food for children (8);
- promoting growth of Lactobacillus bifidus (9);
- improvement of flow of slurries such as cement (10);
- cryo-protectant for storage of organs, tissue and cells (11, 12, 13, 14,);
- coating of tablets (15);
- building block for chemical specialties.

The use of raffinose in the preparation of the artificial blood "Hemolink" is well known. From out of date blood the hemoglobin is extracted. Next the antigens and other proteins are removed and the free hemoglobin is cross-linked with a raffinose derivative. Hemolink can circulate in the body for several days before the body removes it. It can be used for all types of blood. The raffinose derivative is prepared by oxidation of raffinose with periodic acid at pH 5-7. To prevent hydrolysis the product is stored at the same pH. By reacting the resulting polyaldehydes with hemoglobin tetramers a Schiff base is obtained which is then reduced to a stable compound (16, 17).

Raffinose can also be oxidized with galactose oxidase. This enzyme is able to oxidize the C_6H_2OH group in galactose into an aldehyde. The resulted aldehyde sugar can be used for various conversions such as a reaction with free amino groups (18).

Raffinose and various other sugars can be converted into acrylic esters with vinylacrylate and hydrolytic enzymes such as amylases, lipases and proteases. Polymerization of these esters results into poly-sugar acrylates. Copolymerization with 2-hydroxy methacrylate is an option. Poly-sugar acrylates give hydrogels which can be applied in hygienic pads, packaging and drug delivery polymers (19).

Fatty acid esters of raffinose with a carbon chain length of 14 to 20 carbon atoms can be used in cosmetic formulations. They are hypo-allergenic and have moisturizing properties (20).

Hydrogenolysis of raffinose could probably result into polyols with interesting properties for the synthesis of polyurethanes, alkyd resins etc.

Tetrachloro raffinose can be used in the preparation of sucralose. It is prepared by reacting raffinose with thionylchloride in the presence of trifenylfosfineoxide. The 6-chloro-deoxygalactosyl group can be removed with special enzymes (21).

Lactosucrose

Some other small carbohydrate molecules such as stachiose and erlose are difficult to digest and are converted by micro organisms in the colon. On one side this results into flatulence but there is also a stimulation of the intestinal flora. This fact led to the synthesis of various new oligosaccharides/syrups based on cheap sugars such as lactose, saccharose and starch hydrolysates. In a patent of Hayashibara (22) the preparation of lactosucrose (4(G)-β-D-galactosylsucrose) is described starting from a mixture of lactose and sucrose. By using suitable fructosyl or galactosyl transferases a sugar mixture with at least 45% lactosucrose is obtained. After separation and spray drying a powder with a high content of lactosucrose is obtained.

Lactosucrose is an isomer of raffinose. The difference is the way in which the galactose is coupled to the sucrose, β 1→4instead of α 1→6

Oligosaccharide phosphates

Potato starch contains a small amount of phosphate, bound as ester with a DS of 0.002-0.005.The phosphate groups are bound to the B-chains of the amylopectin molecules. About 60%-70% of the phosphate groups are bound to the C_6 of the glucose residues and the remainder to the C_3. The phosphate groups hinder the enzymatic degradation of the starch. This is the reason why oligosaccharides with phosphate groups can be prepared from potato starch by conversion with β-amylase, glucoamylase and pullulanase. The different sugars are separated by column chromatography. Two fractions are obtained in this way. One fraction contains maltotriose, maltotetraose and maltopentaose with *one* phosphate group per molecule. The other fraction contains maltopentaose and maltohexaose with at least *two* phosphate groups per molecule. This latter fraction inhibits the in vitro precipitation of calcium phosphates, an important fact in the bio availability of calcium. The precipitation of calcium phosphate prevents the uptake of calcium in the intestine (23). When preparing maltodextrins for baby-food the content of phosphated oligosaccharides should be low. This is realized by applying phosphatases such as fytase.

Starch can be phosphorylated rather easily according to Neukom's method by heating starch with a mixture of mono and di sodium phosphate at 155 °C during several hours (24, 25). The low substituted Neukom phosphates show mainly C_6 substitution (26). Starting with these potato starch phosphates it might be possible to synthesize specific oligosaccharides with a high phosphate content and more effective in inhibiting the precipitating of calcium. A Japanese patent (27) describes the preparation of phosphorylated sugars based on potato starch with hydrolases and transferases (CGT). Phosphorylated guar hydrolysates also show a calcium phosphate precipitation inhibiting effect (28).

The Neukom phosphorylation of starch has been known for a long time, the same reaction with sugars is used only since the last ten years. After freeze-drying of a sodium phosphate buffered (pH 5.5) aqueous solution of trehalose the dried product was kept at a temperature of 56 °C for several hours. After analysis the presence of four isomeric trehalose orthophosphates was found (29, 30). Also examples of the phosphorylation of sugars such as sucrose and lactose are described (30). It should be possible to prepare raffinose phosphates in the same way.

Glycerol-1-phosphate can be prepared enzymatically. An industrial process for this product has been developed. In this process an alkaline phosphatase immobilized on corn grits is used under reverse conditions. The reaction is highly region selective for the 1- position of the glycerol. This method is maybe also possible for the phosphorylation of raffinose and other oligosaccharides.

Various

In the past various starch ethers and esters have been hydrolyzed enzymatically or with acid. The application of hydrolyzed hydroxypropyl starch for use in sugar free confectionary has been patented (31). The isolation of specific oligosaccharides from such hydrolysates is not simple.

Pectin contains ferulic acid ester groups. By acid or enzymatic hydrolysis oligosaccharides with ferulic acid esters groups can be prepared (32), see chapter on ferulic acid.

Simple oligosaccharides such as maltotriose, maltotetraose, maltopentaose etc. can be relatively easily prepared with suitable enzymes. Use of amylases from Pseudomonas stutzerie in the hydrolysis of starch results in mainly maltotetraose. A process for large scale production has been developed using an immobilized enzyme bio reactor system (33).

References

1. Chem. Magazine. 1987; November: 701.
2. S. Koujiet et al., Zuckerindustrie. 1992; 117(11): 894-899.
3. USP 6,406,547 (2002).
4. USP 6,440,222 (2002).
5. USP 6,224,684 (2001).
6. JP 01137991 (1989).
7. World J. Microbiol. Biotechnol. 1997; 13: 583-585.
8. USP 6,156,368 (2000).
9. USP 6,451,584 (2002).
10. USP 6,478,870 (2002).
11. USP 6,187,529 (2001).
12. USP 6,060,232 (2000).

13. USP 5,827,640 (1998).

14. USP 5,821,045 (1998).

15. USP 6,451,344 (2002).

16. C&EN. 1996; 74: 26.

17. USP 5,532,352 (1996).

18. R. Schoevaart et al., Carbohydr. Res. 2001; 334: 1-6.

19. USP 5,474,915 (1995).

20. USP 4,699,930 (1987).

21. USP 4,826,962 (1989).

22. USP 5,296,473 (1994).

23. H. Kamasaka et al., Biosci. Biotech. Biochem. 1995; 59(8): 1412-1416.

24. USP 2,865,762 (1958).

25. USP 2,884,412 (1958).

26. R.E. Gramera et al., Cereal Chem. 1996; 43: 104.

27. JP 08104696 (1996).

28. O. Watanabe et al., Biosci. Biotech. Biochem. 2001; 65(3): 613-618.

29. E.Tarelli, S.F. Wheeler, Chem & Ind. 1993; 164.

30. WO 94/02495.

31. EP 730828 (1996) Cerestar.

32. M.C. Ralet, Carbohydr. Res. 1994; 263: 227-241.

33. T. Kimura et al., Starch/Stärke. 1990; 42: 151.

Printed in the United States
by Baker & Taylor Publisher Services